Elevated

Cannabis as a Tool for Mind Enhancement

Sebastián Marincolo

Praise for Sebastián Marincolo's work

"I agree with Marincolo's argument about the connection between Benjamin's meditations on hashish and his meditations on art in the "Work of Art" essay and elsewhere. I believe, as Marincolo suggests, that there are a good many connections with other areas of Benjamin's thought as well"

– **Howard Eiland**, lecturer at MIT School of Humanities, Arts and Social Sciences, co-editor of *Benjamin's Selected Writings* and author of a biography on Walter Benjamin, on *What Hashish Did to Walter Benjamin*, Khargala Press, Stuttgart, 2015

"I'm loving it! Wow. What an amazing, important work! Your book is such a treasure – a unique resource. Such a definitive statement of everything I've been thinking when it comes to marijuana … I now understand the benefits that Norman Mailer, Carl Sagan, and Richard Feynman got from Marijuana."

– **Jason Silva**, storyteller, futurist, keynote speaker, known for hosting National Geographic's Brain Games, on *HIGH: Insights on Marijuana*, Dog Ear Publishing, Minneapolis, 2010

"Cannabis is one of earth's power plants on earth, and Sebastian's book demonstrates how to use it skillfully as a performance enhancer that can help us all understand the states of our own minds. Highly recommended for aspiring and experienced psychonauts."

– **Joe Dolce**, author, *Brave New Weed* and host, Brave New Weed podcast Founder, Medical Cannabis Mentor, on *What Hashish Did to Walter Benjamin*, Khargala Press, Stuttgart, 2015

"Marincolo's writing and manner of thinking are excellent – I am enthusiastic about recommending his work."

– **Michael Backes**, author of *Cannabis Pharmacy – The Practical Guide to Medical Marijuana*, on *HIGH: Insights on Marijuana*, Dog Ear Publishing, Minneapolis, 2010

Elevated

Cannabis
as a Tool for
Mind Enhancement

Sebastián Marincolo

HILARITAS
PRESS

Elevated
Cannabis as a Tool for Mind Enhancement

International Standard Book Number (Print): 978-1-952746-21-5
International Standard Book Number (eBook): 978-1-952746-22-2

First Edition 2023

Photographs by Sebastián Marincolo

Cover Design by amoeba

Book Design by Pelorian Digital

Hilaritas Press, LLC.
P.O. Box 1153
Grand Junction, Colorado 81502
www.hilaritaspress.com

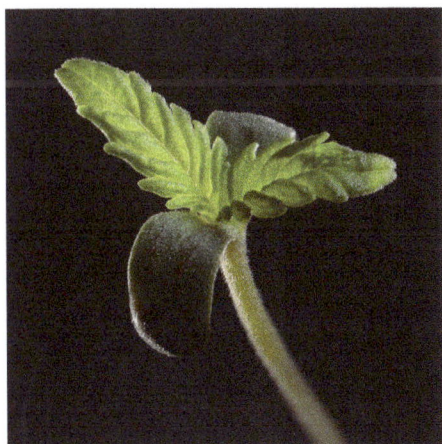

Contents

Foreword

By R. Michael Johnson

Sometime in the late 1970s or early 1980s, I was a lanky, very long-haired, ultra-bookish teenaged heavy metal guitarist in the San Gabriel Valley suburbs of Los Angeles who'd often get stoned with bandmates after practice. Sometimes even *before* we practiced, and other times, *during*, but anyway: I distinctly remember an ebullient group conversation among all of us. Maybe it was a series of bull sessions; the memory is by now not perfect. We were talking about how good the "new" cannabis from Northern California was (at that point, it was all "sensemilla" – Spanish for "no seeds", not "Purple Urkel" or "Girl Scout Cookies" or "Purple Kush" or "Blue Dream"), and we began testifying impromptu to all the wonderful things we got from this herb: remembering incidents we thought we'd forgotten, how sex with our girlfriends was on "some other level" now (at least one of us said such), how Coltrane and Bach and Ustad Shujaat Khan suddenly made more "sense" and that they were perhaps the apex of human musical creation even though at the time we were playing copies of Led Zeppelin and Judas Priest and Ozzy; that when we drank alcohol we got rowdy and mean and sloppy and stupid and felt bad the next day, but one guy said he got high during lunch time at high school and went back on campus and had all kinds of empathy for that one weird kid that everyone picked on. Maybe he acted on that empathy. I don't recall. And yeah, all that made sense to hear. I said that even a glass of tap water the other night tasted divine when I was high. I think they laughed, but it was true; it's happened to me many times since: a slice of plain sourdough bread and a glass of water from the tap seemed miraculous, like *manna*. It sounds kooky . . . unless you've "been there." I could go on and on with remembered conversations like this, but at some point I definitely did say to my friends that "There should be a book about all this."

Forty some-odd years later, you're holding it in your hands. Finally! This is a unicorn; it's the first book of its kind. Sebastián Marincolo breaks the ice by addressing the Final Frontier of Cannabis: its ability to enhance our experience of life, and yield insights into ourselves and others.

Why has a book like this – written by a PhD/polymath who takes cannabis as seriously as Einstein did photons, or Darwin did finch's beaks – taken so long to arrive? There's a number of long, depressing answers, but my favorite *short* answer comes from Marincolo's mentor and later friend, the late eminent Harvard psychiatrist Lester Grinspoon, who coined the term "cannabinophobia."

In the late 1960s Grinspoon's best friend, a fellow intellectual, would come over to visit and get high (Lester did not get high at that time) and talk. Grinspoon knew his friend (Carl Sagan, you may have heard of him) was brilliant, but the weed-smoking seemed ill-advised, and Lester set out to write a book on the dangers of cannabis. Immersing himself in the research, Lester soon found out he'd swallowed the pervasive lies the government had been telling everyone since early in the 20th century. As he researched more and more, he found the truth about cannabis was *almost the opposite* of what the perceived "wisdom" was. He'd been a victim of cannabinophobia, and that's Grinspoon's neologism. Feel free to use it. Like the 17th century "tulip mania" in The Netherlands, the witch trials in Salem, the Great Comic Book Scare of the 1950s, McCarthyism, or any number of satanic panics closer to our times (to enumerate but a smidgen of moral panics), this irrational mania has gone on far too long. It's been a long, slow slog overcoming the hurdles of cannabinophobia, from hearing that your favorite musicians or Beatnik writers used it, so maybe it's not that dangerous?, to giggling at Cheech and Chong's antics, to recognizing the health benefits of cannabis, to the seemingly inexorable march of stark, staring sanity of decriminalization or outright legalization for recreational purposes (but mostly for health) in the US. As I write, only four US states remain at zero tolerance for any cannabis, for any reason. My heavy metal teenaged self never would've believed it.

Looking back on all this, me and my crowd never really believed any of the stories that fed the mass mind and kept cannabinophobia going; although I didn't know about this idea then, we were like initiates in some underground gnostic secret society. These days, you probably know some grandma who uses it for some age-related ailment. If you care to dig a little deeper, you'll find that millions of medical cannabis patients with severe medical conditions have legal access to cannabinoid medicine in many countries now. These patients, their relatives, friends and their doctors have marvelous stories to tell about its benefits. It's better this way. As some lame ad once told us, "membership has its privileges," but looking back, it wasn't worth the human suffering of some African-American getting 40 years in Louisiana for being caught with a half-ounce. Screw that noise. And that fascistic idiocy still crops up in the news today.

So anyway: smoking or ingesting cannabis is fun, it can be used for an astonishing spectrum of ailments, you can cook with it, there are many other industrial and other uses for it and it's easy to grow. And there are now a few thousand books – a tidal wave over the past 15 years – covering those topics. But there never was a book about all the *enhancements* the cannabis high can confer. The very types of enhancements my friends and I talked about, with our limited vocabularies surrounding such a topic.

Sebastián Marincolo provides us with this vocabulary. His creation of terminology gives us a new mental space to talk about – and think with – ideas regarding our cannabis experiences. As anthropologists and sociologists have documented well, social reality is created by *talking to each other*. But humans need pegs to hang their ideas about their experiences on. We need the words. We need metaphors. Marincolo delivers the goods here, and this alone makes **Elevated** pull its intellectual weight: episodic memory, synesthesia, isomorphic extraction, which is the found relationship of ideas that have a similarity of structure that you might not have thought of when "straight". The phenomenon of "hyperfocusing", etc. But, perhaps more importantly, he sheds new light on novel dimensions for thinking about things we thought we already knew: about pattern recognition, the mighty triumvirate of food/sex/music (and why those things are *really* good - unto, yea, verily: *ecstatic* - for some us while high). And Marincolo gives us new ways to think about that special human faculty: imagination . . . while high on cannabis.

He urges us that when we decide to get high we might want to *take notes . . .*

These are a few of the delightful things you'll learn about while reading this book. It's one thing to get high and "know" intuitively what he's writing about here; it's quite another to have intellectual tools to bring to bear when you want to communicate these phenomena to others. I think we must consider these sorts of gifts from the intellectual class an overall societal good.

In this, *Elevated* seems in a long line of books that can function as an education for the senses, like Brillat-Savarin's 1825 ***The Physiology of Taste***. Or any book that you not only found interesting and informative, but those works that impel you to experience your reality-tunnel, art, music, nature, science and others' personalities in some non-ordinary fresh way. One of the first things Marincolo advises is to *pay attention* to what you're experiencing when high. It sounds simplistic, but we all know (well, most of us?) we tend to fall back on the mode of "having fun" and just letting things happen as they may, then noticing the high is wearing off, so we may as well go home now. This education of our senses with cannabis is not that cannabis is "doing it to us." It takes *effort*: you are engaging your sensorium in a complex fashion with emotions, signals from your body, and cognition in a synergetic way. It's not a "linear" "first do this, and only then do that and see what your answer is" kind of thing. It's more of a holistic, right-hemisphere way of learning about yourself and your world. Sometimes I think of it as a sort of "yoga": ganja yoga (which I found actually exists!). The Sanskrit word *yoga* meant "to yoke": you're connecting at least two things. In this case, your nervous system and some aspect of the world "out there." It's sensual, pleasurable, interesting intellectually, and sexy. I find this avenue of education thrilling and I suspect or hope that you do, too. If this seems alien to you but you're intrigued, *Elevated* will no doubt prove enlightening.

What makes Dr. Sebastián Marincolo qualified to bring all this stuff down to us? Well, he has a PhD in Philosophy, particularly in the philosophy of mind, that domain of philosophy concerned with the nature of consciousness. The list of his philosophical mentors includes some of the world's most renowned philosophers of our time: Prof. Emer. William Lycan, Prof. Emer. Simon Blackburn, and his German mentor Prof. Emer. Manfred Frank. For his research, Marincolo steeped in the Cognitive Sciences, like neurobiology, evolutionary psychology and linguistics, and analytic philosophy. Add to this, he's impressively well-read in literature, law and history. And, though English is his second language, as you'll see, he writes English in that sort of lucid, pan-European intellectual style that many find refreshing in its cosmopolitan tone.

When I wrote above that this is the first of its kind, that it's a "unicorn": clearly, there have been individuals who could have written that book I wanted to read so urgently when I was 16 or 17, so why didn't they? Carl Sagan could have (read his florid testimony in Lester Grinspoon's *Marijuana Reconsidered*: Sagan is "Mr. X" . . . or find it online; it's all over the place), but perhaps his adoring public wasn't ready to understand that the charming, spell-binding explainer of the Cosmos was a pothead. Alas, it would've been a different book, too, as Sagan did not have the deep background in modern philosophical thought Marincolo has. Sagan's best friend Grinspoon was most definitely heading toward this territory. So, Grinspoon was delighted when in 2008, Sebastián contacted him by email. Grinspoon asked Marincolo to get on the phone and after one hour of talking in their first phone call, he suggested to co-edit a book with him, a selection of anecdotes and essays Grinspoon had been collecting on his website project marijuana-uses.com.

Lester would be proud to have seen his acolyte and friend Sebastián's new book, this one you're holding now. He read the earlier, much shorter German version of this book – the present work has three added chapters and the other chapters all have been revised and updated – after years of conversations with Sebastián, and wrote the foreword for the German edition that appeared in 2013 and helped make it a big success in Deutschland. The German edition includes a powerful political essay; its big media impact arguably helped to shift the public's opinion – in 2017, Germany introduced a new law for much better access to medical cannabis and recently, the German Government has announced that it will legalize cannabis for adult use very soon.

Lester thought it was important that famous people, especially intellectuals, come out of the cannabis closet and serve as examples in an effort to overcome cannabinophobia, but Sagan apparently saw it as too much of a gamble. Cannabinophobia strikes again!

Other academics have come out of the closet as cannabis enthusiasts over the past forty years, but not many, and they're not loudly flaunting it. You'd better be well-established and have tenure first. "Radical" academics are conservative that way. Chalk it up to cannabinophobia? We're more likely to find informed, eager and spirited defenses of cannabis from that vast crowd of non-academic intellectuals, poets, musicians, painters, actors, and comedians. (Any scholar of Humor should be interested in the juxtaposition of two odd things that creates surprise and a laugh, with Marincolo's ideas about cannabis and "isomorphic extraction.") When these folk do wax on about how wonderful and beneficial cannabis has been for their creative process, or its ability to expand their empathy, or gain new insights about how they see themselves, etc, they don't do it from the vantage point of a thinker with the sophisticated training Marincolo has. We hope he's not the last well-trained intellectual to concentrate his or her thinking on the phenomena of cannabis's various enhancements. Which brings up two key issues.

First, when I initially read this manuscript, I was relieved to find a level of nuance that seems woefully rare when reading about cannabis. All too often the writer seems either convinced this drug is a slippery slope and it's only a matter of time before the user ends up sitting around the house all day on heroin or meth. Or, just as irritating, everything about cannabis is sheer magic: it's a panacea, and if everyone got stoned, the world would be healed. I exaggerate here, but only a little. Websites seem particularly guilty of this, and it's not difficult to discern the "Green Rush" money-grab in a lot of writing about cannabis on the Internet. Marincolo is careful and nuanced at all times. He's not out to make everyone a stoner. You'll notice that early in the book he mentions the time he got horrifyingly way too high. He knows about such things and is candid, frank, and can explain what to do when you're too high. Also, he knows of the vast eccentricities of each individual nervous system and the sets and settings people find themselves in. He's writing with the possible variations and variabilities in the phenomenon of getting high always in mind. It's simply not for everybody. Nor *should* it be. How refreshing!

Secondly, from a precociously early age he was interested in what he'd later learn was famously labeled as "the hard problem" by philosopher David Chalmers: "consciousness": how do we account for physical phenomena like processes in the brain, with experience and our mental states, and how do I know that you experience things like I do? Is *qualia* real, and to what extent?, etc. As he matured in his studies, and continued to be fascinated by his own experiences while high, he thought that cannabis would be an interesting topic to study in the philosophy of consciousness and "Mind." This is where he ran into trouble. Not trouble with cannabis or the welter of insights he found, but cannabinophobia in the Academy. His personal stories are maddening, and you'll get a taste of some of it in this text. He's suffered professionally due to his expertise in his subject. It's a damned shame. All I could say – and I fervently believe this to be true; I hope I'm not wrong – is he's ahead of his time, and the culture just has to catch up to him. He has blazed the trail here.

Some areas of the world are ready to receive this text as joyously as I was, and recognize its vitality; other states, municipalities and prefectures will need more time. Again, you'll make your own decision about this for yourself, as you read.

Finding Sebastian Marincolo was a very happy accident for me. Of course, I wanted to read a book on his topic and was always searching. There were some really fine books that came close, but no one who could write about cannabis with his level of skill and particular training. I read Jack Herrer as soon as his book came out. I loved Martin Lee's history, *Smoke Signals*. There was a time when I'd read every book in my local library that had anything to do with cannabis. There were always fantastic passages about cannabis in Terence McKenna, William S. Burroughs, and Robert Anton Wilson, among others.

One night, I was looking through a large library database about writers who were part of the Frankfurt School of critical theory, and I was also very interested in writers I considered "para-Frankfurt", like Norman O. Brown, Karl Mannheim, and Walter Benjamin, among others. Up popped a book I'd never heard of: ***What Hashish Did To Walter Benjamin (2015)***, by Sebastián Marincolo. I'd never heard of the author, but *mein Gott!* What a title! No one near me had a copy, so I found a copy to buy online. It was as close to my 17 year old self's desire: many essays on various aspects of cannabis and how it enhanced different users in history, not just Benjamin. He's writing about Iain McGilchrist's ideas about the right hemisphere of the brain and what Carl Sagan waxed about so eloquently when he was high. The French Club du Hashischins in the 19th century; jazz and Mezz Mezzrow; the German philosopher Ernst Bloch, who sat down in the roaring 1920s in Berlin with his friend Walter Benjamin, experimenting with Hashish to attain philosophical insights; the names of Antonio Damasio, V.S. Ramachandran, the Blakeslees and interoception scattered throughout the text, etc: this was more like it!

The new edition of *Elevated* can be seen as the up-to-date intellectual foundation for readers to understand in how many ways cannabis can enhance individual minds. In "What Hashish Did To Walter Benjamin", Marincolo adds perspectives on many more enhancements and he then uses the results of his research to explain how famous luminaries like Walter Benjamin, Sagan or The Beatles productively used the cannabis high to influence and change the world forever.

I was talking to a friend a couple months ago about the famous-in-the-Sixties lore: the Houseboat Summit at

Alan Watts's place in Sausalito, with all (most) of the late 1960s luminaries there, but basically it's often framed as the Yippies (Abbie Hoffman, Jerry Rubin, Paul Krassner, et.al) and varieties of political activism, vs. Timothy Leary's non-activism: Turn On, Tune In, and Drop Out. The idea was sometimes thought of as materialist street-wise political activism, vs. That's useless: work on yourself to change the world. The society that brought us the Vietnam War is hopeless. I always thought all the dichotomies were false: you can be in the streets AND work on yourself. This cannabis enhancement idea seems like it's all about the individual (you: "Leary"), but never underestimate the power of a room full of you and your friends, all high together. You're high, but all of you high together (you and your friends: "Krassner") seems like a Free Lunch. Like it ends up being more than the sum of its parts. Something anti-entropic. Then again, you hope everyone gets home safe. Why doesn't everyone just sleep here tonight?

So, following Robert Anton Wilson's writer-friend David Jay Brown's advice he got from *his* friend, computer philosopher Francis Jeffrey: Brown wrote that Jeffrey "said that whenever I enjoy reading a book I should contact the author and tell him or her how much I liked it. He said I would be surprised how many authors would be happy to hear from me." (***The New Science of Psychedelics***, Brown, p.23) It was easy to find Sebastián online: he had a beautiful website with lots of his own gorgeous photography of cannabis plants, buds, etc. I wrote him a fan letter, and he wrote back. We began corresponding and I found him not only seemingly otherworldly in his expertise, but very warm, hilarious, and modest. Every time I asked him a question that had anything to do with cannabis he was spot-on. He told me of books I'd missed and why they're worthy of reading: mostly due to people *writing or talking about their experiences* on cannabis, which is underrated. A few stories are informing, but thousands become a bona-fide data-set. He told me of William Novak's **High Culture: Marijuana In The Lives of Americans** (1980) and Erich Goode's **The Marijuana Smokers** (1971), both of which are still excellent sources, and ones I'd missed. He writes of vaporizing tech, how different burn temperatures can affect cannabis experience, and that whole amazing world of the endocannabinoid system and healing, Prof. Raphael Mechoulam's research, terpenoids and flavonoids, etc. A leitmotif that runs through Marincolos's work is that government cops should get out of the way and *let scientists do the research!* There is so much more to learn about this amazing plant!

I've done a thought experiment a few times: who in the world could know more than Sebastián Marincolo about the enhancements offered by cannabis? Maybe some small group of PhDs who were hedonic, experienced and lovers of cannabis, with unlimited funding for research. I doubt this scenario exists. Sebastián Marincolo and his **Elevated** is probably the best you can possibly do at this moment in history.

– R. Michael Johnson,
Penngrove, California
5 October, 2022

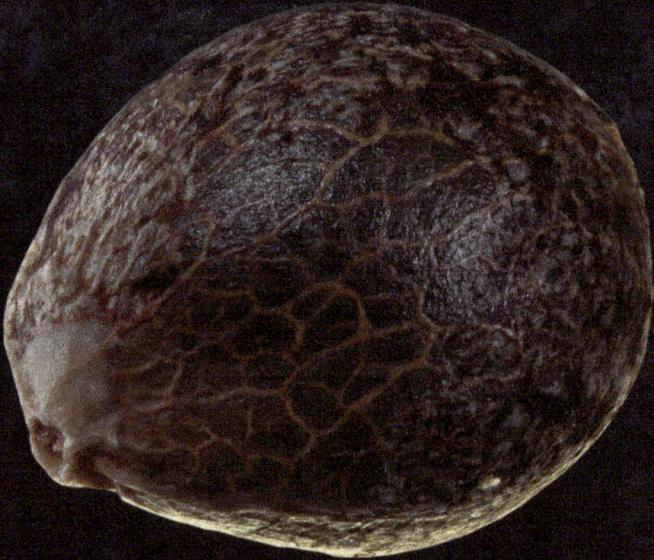

(1) A cannabis seed of the White Haze variety, approximately the size of a match head.

Introduction

"All good science begins as an imaginative excursion into what might be true."
— Peter Medawar, Nobel Prize winning biologist

Twenty-five years ago, I was in my late twenties and working on my doctoral thesis in philosophy. Since my early childhood, I had been interested in the human mind, and I vividly remember how thrilled I was to discover sometimes during my adolescence that there were scientific approaches that exclusively dealt with the questions I had raised very early in my life. My childish mind was still full of questions, and I had never given up asking them. How can a human brain "produce" consciousness? What kinds of animals have consciousness? Elephants, for sure. How about mice? Grasshoppers? Do even they have experiences, or what we'd call a subjective point of view? Can computers develop consciousness? What kinds of systems can support conscious experiences?

At the age of 16, I was already determined to study philosophy, and, specifically, the philosophy of mind, cognitive sciences, and other related disciplines like neuroscience, linguistics, and evolutionary psychology to investigate the phenomenon of consciousness – human consciousness as well as animal consciousness and the possibility of artificial consciousness. I followed my passion and was set straight for an academic career in philosophy. Some years later I was lucky to be able to study philosophy and linguistics at the German Eberhard Karls University of Tübingen, founded in 1477.

My university enabled me to go for a year of academic exchange to the University of North Carolina at Chapel Hill in 1993. I returned to Chapel Hill for another year in 1997 to conduct research for my German doctoral thesis. Around that time, I started again to use cannabis to experiment with getting high. I had already tried it a few times in my early twenties but had then stopped using it after eating a hashish praline which had left me glued to my chair for an eternal hour, with fragmented short-term memory, and a horrific state of disorientation and fear. In 1997, when I tried cannabis again, I approached dosing more mindfully. And this time, my scientific interest was fully sparked. I started to discuss the effects of the cannabis high on my mind with an American friend who studied toxicology at the time. He, also, was stunned by the manifold psychoactive effects of cannabis on his mind. We started to discuss these effects as we both experienced very similar states, but approached descriptions of them from different scientific angles. Both of us experienced many mind enhancements during this altered state of consciousness.

We felt elevated.

The Alice-in Wonderland Route to Understanding Human Consciousness

I was amazed by the multidimensionality of the elevated state of a cannabis high and its temporary effects on perception and many higher cognitive processes and abilities, including various functions of my memory, attention, pattern recognition, imagination, creativity, introspection, as well as on my ability to come up with deep and spontaneous insights and empathic understanding. We both found that while sometimes, some of those negative temporary effects on the mind would make it rather difficult to follow complex thoughts, like the effect on short-term memory ("losing the thread"); we experienced many other effects as enhancements of our consciousness. Phenomena like mind racing lead to hilarious brainstorming processes, and our enhanced pattern recognition led to unusually deep conversations. Our cannabis highs led to conversations and explorations propelled our friendship and created a strong bond between us.

"Wonder is the feeling of a philosopher, and philosophy begins in wonder," says Socrates in Plato's dialogue Theaetetus. I had begun to wonder. In the late 1990s, I decided to start investigating the cannabis high after completing my dissertation thesis. Later, I would call this enterprise the *"Alice-in-Wonderland Route"* to research human consciousness. I knew that researching the human high would be a very special and efficient way to find out more about the nature of consciousness. My research was always about the human mind, not only about the cannabis high as an altered state of consciousness. I used the high as a self-induced altered state of consciousness that enabled me to get insights into the phenomenon of human consciousness itself.

In his book, ***How to Change Your Mind. The New Science of Psychedelics***, author Michael Pollan describes the renaissance of psychedelic research in the last few years and states:

"Today after several decades of suppression and neglect, psychedelics are having a renaissance. A new generation of scientists, many of them inspired by their own personal experience of the compounds, are testing

their potential to heal mental illnesses such as depression, anxiety, trauma, and addiction. Other scientists are using psychedelics in conjunction with new brain-imaging tools to explore the links between brain and mind, hoping to unravel some of the mysteries of consciousness. One good way to understand a complex system is to disturb it and then see what happens."

Let me add to this that I believe that many researchers have underestimated the cannabis high, its mind enhancements, and its potential for researching human consciousness – even many of those researchers belonging to this group of psychedelic researchers.

As my attentional focus changed during a cannabis high, I realized the importance of attention control in our thinking and life. I learned the important connection between episodic memory retrieval and empathic understanding when a high helped me to retrieve memories about a distant relationship breakup, which in turn helped me to better understand how a friend of mine felt in a similar situation. My altered pattern recognition during a high showed me in how many ways we have to be able to recognize patterns in order to function as humans. Studying altered states of consciousness can teach us a lot about human cognition and consciousness – about our very nature as conscious beings.

Tragically, two world wars with myriad soldiers suffering from various brain injuries have helped scientists to better understand the cognitive architecture of the brain by seeing so many partial breakdowns of cognition affecting other cognitive and perceptual abilities in complex, yet systematic ways. The advantage of studying a psychoactive substance for a researcher like me, of course, over studying traumatized patients is that one can experiment upon oneself and observe the phenomenology of altered states of consciousness firsthand.

Cannabis, Taboos, and Rationality

After completing my German doctorate in philosophy in the field of the philosophy of mind and neurocognition I started to research the cannabis high, as I had planned. In retrospect, I have to smile at my past self (forgivingly) for even trying to get research grants for several years to investigate the positive mind-altering potential of the cannabis high. Of course, all my requests were denied, and there were only a few organizations worldwide to consider. That was in 2004. I had finally understood for good that my decision to research the cannabis high had wrecked my academic career – so I took a different path. To finance my research, I jumped into a rabbit hole of life, working as a photographer, creative director, PR consultant, and later as a Director of Communications and Marketing in the medical cannabis industry. I do not regret my decision. Studying the cannabis high has helped me to come to a much deeper understanding of consciousness and the core abilities that make us humans so special, abilities such as imagination, creativity, introspection, empathy, and our ability to produce deep insights.

I knew I was dealing with a taboo subject when I started my research. I was not aware, however, of the full negative impact of this cultural taboo on so many when it comes to simple reasoning. Even liberal, intelligent, and highly educated people stop thinking and arguing rationally when it comes to the discussion of the topic of cannabis – especially when it comes to the aspect of its positive mind-altering potential. I have come to understand that many people do not care about good arguments. If you accept and defend an opinion that is not part of the mainstream, and maybe even considered a taboo position to hold, then it is just easier not to become convinced. You want to be on the winner's side.

As a researcher, you get a lot more understanding these days if you research the medicinal aspects of cannabis, or the potential of cannabis as a useful plant for nutritional or industrial purposes. When addressing these aspects, the taboo concerning the cannabis plant has already eroded to a significant degree in many countries in the last decades and cannabis has been re-discovered worldwide in various regions for its medicinal uses. Many of these medical uses had of course been discovered and were documented thousands of years ago in various ancient cultures. The mythical Chinese Emperor Shen-Nun, considered by many as the father of Chinese medicine, allegedly recommended the female Cannabis plants for menstrual fatigue, gout, rheumatism, malaria, constipation, and "absent-mindedness". While a written record of that time does not exist and historical data cannot establish the existence of Shen Nun around 2000 B.C. it seems that we can safely say that medical cannabis use in China probably started in shamanic times. Many other ancient pharmacopeias from India to Egypt and Greece list cannabis as an important therapeutic option for a whole variety of other medical conditions.

Modern research has confirmed to a great extent that cannabis can be a relatively safe and effective treatment for a wide range of medical conditions. The discovery of the human endocannabinoid system in the 1990s has given us a much deeper understanding of the physiological mechanisms that mediate the effects of plant-based cannabinoids in humans. Most experts today believe that the system is involved in the regulation of stress, sleep, pain, memory, learning, the immune system, and many others functions. (I will give an introduction to the Endocannabinoid System in chapter VIII "Mind Enhancements and the Endocannabinoid System".)

In November 2022, 37 States in the U.S. and more than 50 countries worldwide allow for some kind of access for patients to medical cannabis. Many recent studies and the experiences of health care practitioners and hundreds of thousands of cannabis patients in the last years indicate that medical cannabis can be successfully used for the symptomatic treatment of chronic pain, neuropathic pain, nausea, spasms, lack of appetite, glaucoma, asthma, epilepsy, depression, Tourette's syndrome, and many other medical applications.

Hemp products like cannabis chocolate, hemp-seed oil, and hemp clothing conquer stores worldwide. Hemp fibers are extremely strong and are used as an insulating material for construction. Car companies produce various composite plastic parts with hemp fibers, replacing the synthetic fibers used before. Hemp products are used for many other purposes, including the production of textiles, cosmetic articles, or building houses. The rediscovery of industrial hemp is in full swing. The market for hemp-derived cannabidiol (CBD) was already at an estimated 2.8 billion USD worldwide in 2020 and will grow further.[1]

Many of those who have so fervently upheld the prohibition of cannabis for such a long time have changed their minds. For some, mostly under economic pressure; for others it looks like it was peer pressure. In some regions of the world, opposing cannabis has become very unfashionable. While people have been truly convinced by rational arguments that cannabis prohibition makes no sense and that we should explore its potential, others want to go with the flow now. They want to be on the winner's side again.

Cannabis and hemp are becoming hip – or so you would think. When it comes to the positive enhancement potential of the cannabis high as an altered state of consciousness, old myths and taboos based on decades of disinformation campaigns still prevail in many countries. In some regions – some states of the U.S., Canada, Uruguay, or Jamaica – entrepreneurs of various strands have started to explore the usefulness of a cannabis high for yoga, sex, or meditation. Others still consider the cannabis high as the entry gates to the hell of addiction and insanity.

The ignorance of so many reports about cannabis being a mind-enhancing substance in our culture today not only seems to stem from fabricated myths of past disinformation campaigns. The roots of this ignorance are much deeper. Generally, we need to understand that many of the reported mind-enhancements do not fit the contemporary Western stereotype of a mind enhancer such as amphetamines, which mainly help you to stay awake and focused for a long time, often at the expense of attending to empathic signals.

As Lester Grinspoon's best friend and avid cannabis user Carl Sagan observed in a footnote in his Pulitzer Price-winning book *The Dragons of Eden*, cannabis seems to enhance many cognitive brain functions we usually see as being mainly right-hemisphere-based, such as episodic memory, imagination, pattern recognition, introspection, creativity, and empathic understanding. In his eminent book *Marijuana Reconsidered*, Lester Grinspoon points out:

"(...) (C)annabis has been accepted for centuries

among those people in India where cultural background and religious teaching support introspection, meditation, and bodily passivity. The West, with its cultural emphasis on achievement, activity, and aggressiveness, has elected alcohol as its acceptable, semiofficial euphoriant. (...) The more introspective, meditative, nonaggressive stereotype associated with marijuana goes against the Western cultural mainstream, particularly in the United States. "[2]

Certainly, there are risks and potential for abuse when it comes to cannabis. These risks should be taken seriously. Yet, I am convinced that there are compelling reasons to legalize and regulate cannabis for responsible adult use, comparable to the way we regulate the drug alcohol. Many scientific studies have shown that this is a better way to keep consumers – and especially adolescents – from abusing the substance. In this book, I will mostly address the positive potential of the cannabis high, but this certainly does not imply that I want to negate possible risks or deny the fact that cannabis abuse exists. Rather, I would like to concentrate on the phenomenon of the cannabis high and its positive potential because the worldwide prohibition in the last decades did not allow for much research concerning this topic – and I am convinced there is a desperate need for more research in this area.

Voyage Into The High

How do you research an altered state of consciousness like the elevated state of a cannabis high? After making my own experiences, my research included a critical analysis of hundreds of detailed reports from various sources, some of them quite recent, while others came from older literary reports. The French poet and writer Charles Baudelaire, for instance, had founded a 'Hasheesh Club' in Paris with other important intellectuals and writers of his time to eat hashish marmalade and write about the cannabis high. In 2008, I contacted Harvard Associate Prof. Emeritus Lester Grinspoon, one of the most prominent experts on medical marijuana worldwide. I had long been interested in his research and vividly remember how stunned I was when I began to read the many detailed reports and essays about what he would call the enhancement uses of cannabis, collected on his groundbreaking website marijuana-uses. com. It was so reassuring to see how these literary and other reports from cannabis users with various cultural backgrounds confirmed to me to a large degree the experiences I had made myself with the high: those very experiences which I had discussed with my American toxicologist friend long before I started to investigate the cannabis high scientifically.

The essays and reports about the cannabis high collected by Lester Grinspoon[3] as well as many more detailed anecdotal reports collected by other authors such as William Novak,[4] Charles Tart,[5] or Erich Goode[6] describe an astonishing spectrum of cognitive effects during a cannabis high. Many users experience an enhanced ability to vividly remember past events in amazing detail, they experience hyperfocus of attention, their sensations become more intense, and they often experience them with more complexity and detail in structure. They report an enhanced imagination, a heightened perception of more nuances in music or art, or become more creative when cooking or in sexual encounters. They generally discover new patterns of which they were previously unaware – visual patterns in star maps, acoustic patterns in music, or complex behavioral patterns in human actions. Many users report an enhanced ability for introspection or tell us how a cannabis high helped them to better empathically understand others. And many users report spontaneous, deep insights, which they often evaluate as positively life-changing and incredibly useful for their personal development.

Lester Grinspoon wisely pointed out that these anecdotes have to be approached critically. We need to be extremely careful to not to jump to conclusions. However, he also emphasizes that the successful use of medical cannabis in recent years rests to a significant degree on anecdotal evidence from users and reminds us that anecdotal evidence has generally played a major role in the development and modification of medications in our history.

Anecdotes from others, however, have always only been one of many sources for my research.

Many people have asked me with a bashful smile whether I had "researched cannabis empirically" myself. Of course, I have. Does this imply that I am biased as a researcher? Certainly, one literary agent who was interested in a previous book project thought so. I refused to erase my own experiences with the cannabis high from my book, and he jumped off.

So, if you think like this literary agent, let me ask you this: If you want to know something about the culture of Tibet, would you rather read the book of an author who never set a foot in the country and only read about its culture? Or would you rather read a book from an author who traveled and lived in the country and also knows books and reports about the country? Surely, if you want to get a more balanced report about all the aspects of the Tibetan culture, you should of course not only read this one book by an author who traveled there, who maybe even shows affection for the country and its inhabitants, but you should also include the opinions, reports, and books of other knowledgeable people. I think it is clear that the author who traveled will know about aspects of the country which an inexperienced person cannot know.

To put it differently, then, I have researched the cannabis high also as a *psychonaut* – the notion was termed by the German writer Ernst Jünger and denotes the act of finding out about one's psyche through the use of mind-altering substances or techniques. My psychonautic journeys were used as another source of knowledge for my overall approach to investigating the nature of the human mind by researching the cannabis high. From a philosophical perspective, this approach is also inspired by the tradition of phenomenology, a philosophical movement founded by philosopher Edmund Husserl in the early years of the 20th century. Phenomenologists approached researching the human mind by systematically analyzing and focusing on the phenomena, the subjective experiences of consciousness.

I integrated the careful and somewhat anthropological analysis of anecdotal reports of the high and my psychonautic approach in a much larger research effort. In my 2015 book **What Hashish did to Walter Benjamin**, I called this approach GUINEA, a *Guerilla-Neurophilosophical Approach* to researching the cannabis high. It is Guerilla because it also integrates knowledge from illegal growers and breeders of cannabis as well as (often) illegal users, as well as from the interdisciplinary field of philosophy of mind, cognitive psychology, neuropsychology, pharmacology, Gestalt psychology, cannabinoid medicine, evolutionary biology, and other related disciplines.

A Few Comments on the Evidential Situation

How do I evaluate the evidential situation for my claims? How strong is evidence so far that a cannabis high can temporarily enhance cognitive activities in the way described in this book?

In my view, even a mountain of anecdotal evidence on experiences of enhanced cognition during a high should not be seen as convincing evidence in itself. Positive anecdotes could be systematically influenced by a bias of users to justify or glorify their use. Likewise, negative anecdotes could be influenced by a negative bias of users or by their inability to control factors that play a role in the successful use of cannabis for creative enhancement. We need to take a very critical attitude towards these anecdotes and ask some questions: Did those who report copy from others? What biases could they have? How did they think about cannabis use before they made their observations, and what were their convictions? How much detail do we find in their descriptions, and can we find complex yet similar patterns in their descriptions of the effects of cannabis on their minds? What do we know about the doses they took and the actual material they consumed? Did they consume other substances, simultaneously, and are their reports only about cannabis?

Then we have to look at the interdisciplinary research of the mind sciences, including neuroscientific theories of the mind, to come to a better understanding of how cognitive functions like creative thinking, pattern recognition, introspection, or empathic understanding are interrelated. If we get a better understanding of how all these cognitive processes are interconnected and what their nature is, we can better access the claims made by anecdotal reports. Of course, we can to some degree draw on the knowledge of medical studies of cannabinoids, patients, the endocannabinoid system, the pharmacology of cannabinoids, the botany of the plant, evolutionary biology, and other scientific resources to come to a better evaluation of anecdotal reports and how credible they are.

Importantly, I believe that the outcome of my scientific approach GUINEA has different implications for the realm of science as opposed to the practical world of everyday cannabis users. Let me comment on science first.

It is important to keep in mind that in science, a hypothesis or theory goes through various stages of gradual confirmation or disconfirmation. Before we start a scientific investigation today we need to compile some initial evidence for a hypothesis, such as anecdotal reports and initial observations. A scientific hypothesis or a theory is not a mere fantasy-construct but can be seen as some initial pattern recognition by experts with some kind of preliminary evidence. And scientific theories, too, have different degrees of empirical confirmation.

We can compare science to our judiciary system. Imagine you are a police officer and come into a room and see a dead man on the ground, with three bullet holes in his chest. Another man is standing next to him with a smoking gun in his hand. According to the law, probably in all countries today, you are now entitled to arrest this man based on your first suspicion. In the U.S., there are various degrees of suspicions that have to be substantiated by increasing levels of evidence on the way to a trial. A "reasonable suspicion" licenses a police officer to pull a recklessly swerving driver over for a field sobriety test, and the more evidence demanding "probable cause" would for instance be needed for an arrest or a search warrant. And, of course, when it comes to a trial later, there are possibilities for an appeal, etc., phases which then involve even more evidence to "prove" somebody guilty or innocent.

I see the evidential situation with many of my hypotheses about cannabis as a mind-enhancing tool somewhere in the league of "probable cause" for scientists today. I wanted to take it way beyond anecdotal evidence and to get a better understanding of how the cannabis high might truly be involved in the various mind enhancements. I certainly do not see this case as closed. My research led me to generate hypotheses about how a cannabis high might positively influence highly complex cognitive abilities like introspection, creativity, or empathic understanding under favorable circumstances. Also, and importantly, I present in this volume my hypothesis that the endocannabinoid system may be involved in supporting functions of the highest cognitive abilities we have. This ongoing research project is one of basic science and I think it has many implications for other scientific disciplines. I am hoping to inspire scientists from various fields such as the cognitive (neuro-) sciences, evolutionary biology, psychology, or the philosophy of mind to investigate this phenomenon. Further research in this area could help us to better understand the role of the endocannabinoid system in human cognition and emotion.

Moreover, I believe that my contribution in this field helps to lead to a better understanding of how and why people use (and abuse) cannabis and in which ways people can personally benefit from a cannabis high. I hope to inspire scientists to further explore the psychoactivity of cannabis to be better able to integrate these effects for medical and many other purposes. My work and the research of Lester Grinspoon and so many others who I refer to should also be highly interesting for health care practitioners working with cannabinoids: psychotherapists, traumatologists, psychoanalysts, and many other professionals.

Practical Implications for Cannabis Users

What should cannabis users take away from my research when it comes to our actual use of cannabis for creativity? Do I recommend that they wait until cognitive scientists have done all that work to answer the questions above? Should we throw our hands in the air in despair because we do not have definitive answers from scientists? Or should we start experimenting for ourselves? What are the practical implications for cannabis users of my analysis so far?

Let me take you on a short and entertaining philosophical detour to prepare an answer to this. Bear with me, I'm sure you will find this interesting.

In his famous philosophical paper "Elusive Knowledge", the eminent philosopher David Lewis convincingly argued that knowledge is a somewhat elusive phenomenon. Knowledge, as odd as that may sound, seems to be context-dependent. In an ordinary context in which I look at my stretched-out hand and claim "This is my hand", my claim would usually be considered knowledge. I just know that this is my hand! Good common sense. But what if I change the context and walk on a stage with a magician who I know is using mirrors to confuse people? When the magician sits me on a chair and asks me to close my eyes and open them again after a few seconds, I may see my hand there, yet will not feel like "I know this is my hand". Suddenly, what seemed to be knowledge to me is questionable.

A long story of prohibition of cannabis and disinformation campaigns that have been ongoing for decades has created a cultural context in which users are faced with a very heavy burden of proof.

And, as one of my philosophical mentors, William Lycan, recently wrote in his book *On Evidence in Philosophy*: *"If you can succeed in placing the burden of proof on your opponent, the game is 99 percent won."*[7]

When we decide to take medicine or to consume some type of food, we all have to make a risk-benefit analysis. I love eating that chocolate bar, but is it worth the fun given the risks of industrial sugar? When concerning cannabis, many have heard countless myths about its alleged danger and toxicity. For a positive risk-benefit analysis, they would need to have a lot of evidence presented to them that cannabis is worth trying.

In my last chapter on the irrationality of cannabis prohibition and addiction, I hope that I can help to debunk at least some myths. Nevertheless, I'd like to stress that this book does not aim to present all the arguments needed for those who want to make a balanced decision about trying cannabis or not. I'd highly advise them to take a close look at the current medical literature on cannabis in order to assess the real risks of cannabis. In my view, there certainly are risks, but I am firmly convinced that these are manageable – and my research certainly gives a better understanding of how to manage the real risks.

For those who are convinced that the risks of cannabis use are manageable and that its toxicity is remarkably low, I think my work has produced significantly more evidence to believe that cannabis can be useful for various mind enhancement purposes. Like many cannabis patients who have concluded that their use helps them for a condition for which there is no conclusive evidence yet, these inspirational users will not wait in vain for stronger evidence from studies. I am hoping to inspire them to make the best of their use of cannabis and to explore the many fascinating ways for mind-enhancing cannabis uses. Many cannabis users reported that their use of this versatile plant deeply and positively transformed their lives and helped them to start a journey of personal development.

(2) A cannabis seed of a different variety. The close-up already shows, compared to the White Haze seed, how different the morphology and color of seeds from cannabis varieties can be. As a plant genus, cannabis can express more than 140 cannabinoids, over 200 terpenes, and dozens of flavonoids. Each variety of cannabis has its own characteristic biochemical profile with characteristic quantities of these substances.

A Science-Art Crossover Project

I reached out to Lester Grinspoon in 2008 during the writing of my first book about the cannabis high, **HIGH: Insights on Marijuana** (Dogear Publishing 2010). Lester answered my email with great enthusiasm and asked me to get on the phone as soon as possible. Since then, we corresponded for more than a decade and discussed the cannabis high in countless conversations over Skype. Lester became a magnificent mentor and a friend to me and I feel truly blessed that I could finally meet him and his wife Betsy in Boston in 2016. His work, his experience, his support and his gracious willingness to share his unbelievable knowledge with me was a true treasure for my research. Sadly, Lester did not live to see this English version of my book **High: Das positive Potential von Marihuana** (*HIGH. The positive Potential of Marijuana*) which originally appeared in 2013 for the renowned publisher Klett-Cotta Tropen in Germany. I think he would have liked to see how much I developed my knowledge in the last years, especially after working for several years in the medical cannabis realm.

Like Lester, I always felt that I have an obligation to influence politics and society for the better. My book, therefore, also contains an essay on the irrational nightmare of the cannabis prohibition, and with Lester's help, I managed to get the word out through TV and radio appearances, magazine interviews, and much international media attention for my work. Germany introduced a new medical cannabis law in 2017, the most innovative law worldwide, giving doctors the ability to prescribe medical cannabis for all kinds of severe illnesses and giving patients the ability to get their cannabis medication reimbursed by their state insurance. In November 2021, the coalition of parties to form the new German government (Social Democrats, Green Party, and the Liberal Party) announced the legalization of cannabis for responsible adult use. Yet, I am truly happy that he could see the positive developments concerning more sensible cannabis policies worldwide in the last years. "The cat is out of the sack", he said to me in one of our last conversations. "Nobody will be able to put it back in".

Most of the chapters in this book are completely revised and updated versions of my German 2013 book ***High: Das positive Potential von Marijuana***. This new edition also includes three more chapters than the original version: Chapter III, *Cannabis, Creativity, and Cognitive Liberty; Chapter VII, Cannabis, Love, Sex, and Tantra, and chapter VIII, Mind Enhancements and the Endocannabinoid System.*

Visual Concept: An Elevated Perspective

The visual concept of the book aims to circumvent learned associations concerning "cannabis", "marijuana", or "hashish," based on earlier disinformation campaigns, which is also why I have abstained from showing the cannabis leaf in its typical symbolic form. My artistic macro photography shows the living plant, the buds, leaves, as well as other plant parts throughout the growth and flowering phases of the cannabis plant. The last photos in the series show tiny parts of the plant which are only a few millimeters in length, making the fine stigmas of the plant look like a strange tree from another planet. The limited art photo series from 2012 shows the fascinating world of the crystalline mushroom-shaped trichomes on the plant's buds and leaves, which produce the psychoactive cannabinoids.

The photo series not only aims to uncover the breathtaking beauty of the plant, it also illustrates the differences in plant morphology of various cannabis strains of cannabis. On a visual level, then, the perceiver can relate to the claim that various strains with their unique combination of dozens of cannabinoids produce significantly different highs – a fact often unknown and underrated by many who see cannabis as a simple drug with only one active ingredient.

The limited photo art series was shot in 2012 with a Canon EOS 5D Mark II and a Canon MP-E 65 macro lens. The photos were generated with a 'deep focus fusion'-technology, which allows for a much greater depth of focus on the extreme macro level. For the backgrounds of the plants, I took macro photos of air bubbles entrapped in ice and color-inverted the photos. My goal was to produce backgrounds giving a more three-dimensional visual experience when looking at the plant, an experience typical for the cannabis high – with strange structures, leaving the perceiver – especially the high perceiver – room for interpretation and pattern recognition.

(See the photograph on page 23)

The essayist Gertrude Stein once wrote in a poem the famous line "A rose is a rose is a rose". She meant to remind us that a name evokes many images and thoughts which are not part of the denoted object but come from our culturally learned associations. Stein concluded that we can only come closer to the object if we manage to cut off these associations, and one trick is to repeat the name all over again.

As opposed to the most romantic associations that arise when we are hearing the name "rose", many of us have mostly negative associations concerning "marijuana", or "hashish", associations that have been drilled into us for many decades by the use of various disinformation campaigns. I wanted to give a new visual perspective along with a new intellectual perspective on the plant which circumvents learned negative associations, a perspective that reminds us that in the end, cannabis is just a plant; and . . .

a plant

is a plant

is a plant.

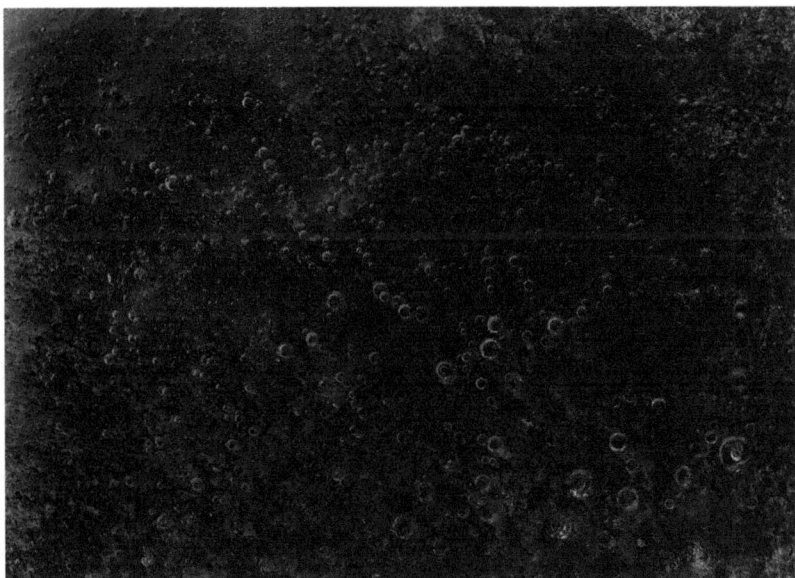

(3) One of the backgrounds I used for my cannabis photography. The backgrounds were produced to increase the impression of an enhanced depth perception, imitating an effect often experienced during a cannabis high. I produced ice layers with trapped air bubbles, used deep focus fusion technology, and then inverted the white photos to black. This background consists of more than 20 layers, the macro photos of the plant add another 20-25 layers.

Editor's note: this photograph was enhanced (brightness increased slightly) so it would display better on non-glossy paper.

(4) A cannabis plant still in the germination phase.

I. Cannabis, Surfing, and How to Ride a High

Imagine yourself in the early nineteen sixties visiting Hawaii for the first time. Walking over the beach you meet a funky guy who shows you an oddly shaped wooden board and tells you to go out and ride the waves. You have heard stories about that thing called "surfing" before. This might be fun! Five minutes later, a big wave throws you to the shore. You try again, but after you have been thrown back on the beach for the third time, you look at your skin rashes and begin to wonder what this is all about.

Would you blame the board for your bad first surfing experience? Obviously, the board was not the problem. You have tried out a wonderful tool that can help you to experience life-changing, blissful experiences – but you need to work on some practical skills first: how to paddle out in the waves, how to get up to stand on a board in the water, how to find your balance and keep it on a wave. Also, you need to acquire a lot of knowledge: which kinds of waves are the best to ride on, and what kind of board should I use for certain situations? Which beaches can be dangerous? Are there undercurrents? And how do I deal with side winds? Importantly, also, you will have to learn to judge your skills: am I good enough to ride this kind of board in this weather? Am I ready yet to ride this tube wave?

Surfboards can be seen as tools in general: they have potential, but to use this potential, we need to learn how to use them and we need some knowledge on what to do best with them. Tools have not only a potential for use but also abuse: I can use a hammer to build a house, but I can also abuse it and hit someone with it on purpose. I can use a car to bring an emergency patient to a hospital and save his life, but I can also abuse it by carelessly speeding on the highway and causing a fatal accident. Tools need skills and knowledge and can be used or abused. Furthermore, even the use of tools brings risks. Even if I use a hammer with good intentions, I can still hit my finger if I have not learned how to safely use it. These risks can be minimized with increasing skills and knowledge, but we usually cannot eliminate risks completely in our life.

All this sounds almost painfully trivial when we talk about tools in general. But we often tend to forget this when it comes to discussions about cannabis and other psychoactive substances. This is a fatal flaw in thinking – with devastating consequences for hundreds of millions of people.

Cannabis as a Tool for Mind Enhancement

Psychoactive cannabis varieties, like any other psychoactive substance, should be seen as a tool. Cannabis and surfboards are tools with a certain positive potential to make great experiences, but also with risks. Our abilities and knowledge will help us to explore their potential and to minimize their risks. Naturally, as for all tools, there is also an *abuse* potential. With the wrong attitude, I can abuse a surfboard as a beginner and try to show off in large waves, endangering myself and other surfers or swimmers close to me. People abuse cannabis for instance when they get high and operate dangerous machinery or drive a car in a condition in which they do not have full control anymore. The cannabis high can be helpful to escape from a harsh reality sometimes and to relieve stress, but under some conditions, this behavior can become a habit that keeps many users from constructively dealing with problems they need to face at some point.

Even if cannabis is remarkably non-toxic and not very addictive in comparison to other psychoactive substances like alcohol or tobacco, it can reinforce an existing need for escapism and, thus, have a negative influence on the lives of those who do not want to confront reality.

On the other hand, millions of people have positively used the effects of cannabis with a different attitude to sometimes concentrate on the here-and-now of existence and thereby relax deeply, explore sensual experiences, or enjoy nature. Your attitude is crucially important as to whether cannabis will affect you positively or negatively.

An unusual attentional focus on the here-and-now or euphoria during a high is only the very beginning of what users have told us about the positive potential of cannabis. Prohibitionists tend to ignore the inspirational potential of cannabis and often reduce its positive effects to a feeling of relaxation coupled with euphoria and stupid behavior. In other words, they see the cannabis high as a happy state of being *dazed and confused*. They would rather concentrate on mostly invented risks and its abuse potential. This prejudiced and uninformed view should not be surprising coming from people who have never used cannabis and have mostly been exposed to disinformation. Sadly, though, even many of those who have tried cannabis share this self-image. Some users have never really used cannabis under favorable conditions, or they cannot remember and conceptualize what a cannabis high has done for them. Many cannabis users in the past have accepted the "standard" negative view on their consumption and feel ashamed to say that they use it for various enhancements because this assertion would just seem like an outrageous effort to justify one's vice. In the past years, I have talked to dozens of cannabis users who confirmed to me that only after reading my essays did they realize how many enhancements they had used with cannabis without being able to verbally communicate it.

Can marijuana lead to an enhancement of creativity? Can a high help us to remember past episodes better, and can we come to important insights about others and ourselves during a high? The answer I will give in the following essay is: yes, definitely, cannabis does have this potential. But, importantly, this is a potential – not an automatism.

Surfboards have typical design features, but they certainly do not let you automatically ride a big wave. A cannabis high brings with it certain typical cognitive effects, like for instance hyperfocus of attention. To positively use these effects, users need skills and knowledge, just like users of a surfboard. They will, for instance, have to know facts about how to choose the right dose, and how to find a variety of cannabis that brings a certain effect profile, because different varieties can lead to different highs. They will also have to learn about various methods of consumption like ingesting or vaporizing cannabis and how different these are in affecting their highs. They should also learn, amongst other things, when they are in the right mood to get high and if certain people joining him/her are a good company to get high with. They will choose their variety, dosage, and consumption method carefully and adjust it to their experiences with riding a high and to their immediate mood, their needs, their activity, and environment, just like skilled surfers will choose the right kind of board matching their skills and the conditions out there in the waves.

If a novice surfer goes out on a Hawaiian beach to ride high waves with a pro board and gets in trouble, he will probably get in a panic. We would not conclude that therefore, his panic is a typical effect of surfing, would we? When a cannabis user panics during a high, it usually happens because he lacks the skills and the knowledge of how, and under what conditions, to ride a cannabis high. It happens to many people, but that does not mean that cannabis always causes panic. What we can learn from this analogy is that a novice and even many experienced users with poor judgment should have more respect, better knowledge, and greater skills before they go out there 'riding a high'.

There is of course a big difference in our society today as it concerns the use of surfboards or cannabis. The globally influential disinformation campaign concerning cannabis started by drug czar Harry Anslinger in the nineteen-thirties in the U.S. invented horror stories about the "evil drug" *marijuana* which are still influential. Let us assume we would run a disinformation campaign for several decades that convinces surfers that most of them will be attacked by sharks and that many others will drown in the waves: how many of them would become paranoid even on a bright sunny, and peaceful day out there on the ocean? And how many more would become paranoid going out into the waves if surfing was strictly prohibited and punished with jail sentences? Now, let us assume we design a study with 10,000 surfers in this situation and, after polling them, conclude that almost 94.7 % of all surfers experience paranoia while surfing. Should we conclude that the surfboards cause paranoia?

In 1965, the American "beat" poet Allen Ginsberg wrote:

". . . most of the horrific effects and disorders described as characteristic of marijuana "intoxication" by the US Federal Treasury Department's Bureau of Narcotics are quite the reverse, precisely traceable back to the effects on consciousness not of the narcotic but of the law and the threatening activities of the US Bureau of Narcotics itself. (...) I myself experience this form of paranoia when I smoke marijuana, and for that reason smoke it in America more rarely than I did in countries where it is legal. "[8]

Ginsberg was right to point out that the cultural context of prohibition has a negative influence on the high of many users. Some years before Ginsberg wrote this, Harvard psychology professor Timothy Leary, later to become famous as the "LSD-guru," had begun to popularize the idea that the character of substance-induced altered states of mind like the cannabis high or an LSD trip depends on three factors: dosage, set and setting of a consumer.[9] The (mind)set includes character, personality, preparation for the experience, intention, mood, expectations, fears, wishes, and attitude of the consumer; whereas the setting includes the context of consumption, the physical environment, but also the emotional/social and cultural environment. [10] Almost a century before Leary conceptualized "set and setting", French writer and poet Charles Baudelaire, a founding member of the mid-nineteenth century Club des Hashashins, had made similar observations in his book "The Artificial Paradises" (1860). After experimenting with very high doses of ingested cannabis marmalade for inspiration, Baudelaire suggested a favorable environment for the hashish high, "like amid a picturesque landscape, or in an artistically decorated room".[11]

The Rediscovery of the Cannabis High

It is becoming increasingly obvious now even to skeptics that cannabis has an almost incredible treatment potential for many medical conditions. But when it comes to the much sought-after inspirational uses of cannabis, it is even more important that we free ourselves of past governmental lies and deceptions and rather carefully listen to what respectful, skilled, and knowledgeable users of cannabis have reported for thousands of years.

This is also the fundamental insight of both the Harvard scholars Lester Grinspoon and Charles Tart more than forty years ago. Psychology professor Charles T. Tart published his study "On being Stoned" in 1971. [12] He had sent 750 questionnaires to students and asked them specifically for their experiences with the effects of marijuana on their consciousness. At the same time, his colleague associate professor for psychiatry Lester Grinspoon meticulously researched anecdotal evidence mostly from experienced cannabis users about their experience of the high in the existing literature and evaluated these reports critically on the background of their situation and their influences and beliefs.[13]

While many feel that they are losing the thread during a conversation and are temporarily struggling with their short-term memory during a high, we have many detailed reports about the enhancement of episodic memory with detailed and vivid recollections of past events that almost feel to subjects like they had been re-living past experiences. Countless users have reported the feeling of awe during the elevated state of a high: they feel a sense of wonder and curiosity and start to attend to whatever comes into the focus of attention with the openness and curiosity of a child. For some, this feeling is the beginning of a life-long journey exploring other styles of music, art, or scientific endeavors. Often, cannabis users have reported that their attention is re-directed and has a different pattern, many of them describe how they attend more and in better detail to their bodily feelings such as being tense or relaxed, they feel their breathing stronger and truly "come back" to their bodies. They experience a whole new dimension and quality in their love life. Cannabis users repeatedly reported a greater acuity in perception, which may have to do, at least partially, with their attentional hyperfocus during a high. Some people use a more refined sense of taste during the high to develop their cooking abilities. A geologist working for the National Aeronautics and Space Administration research center explains how a high helped him to finally merge two stereo photographs into a single three-dimensional view – after he had unsuccessfully tried for months.[14]

Under stronger influences, cannabis users often report synesthetic experiences such as seeing corresponding colors when hearing a guitar solo. Very often, they experience the intensification of their imagination, be it visually, auditorily, gustatory, olfactory, or tactile. And many of them know the phenomenon of mind racing, which some find annoying, while others find it helpful for fast associative mental journeys. Users like the famous astronomer and popularizer of science Carl Sagan, Lester Grinspoon's best friend, and many others have reported enhanced pattern recognition, and most cannabis users know the feeling of a distorted sense of time – sometimes during a high, time seems to move very slowly, which can increase the pleasure of a beautiful or memorable experience. On the other hand, it may sometimes seem as if time just melted away quickly during a high and we are left with a retrospective feeling that a day passed fast-forward. Cannabis users have consistently reported an enhanced ability for introspection, empathic understanding, as well as an enhanced ability to produce remarkable and valuable insights – as well as some stupid ideas during a high.

Interestingly, many cannabis users have reported that the cannabis high can be mood-altering in various ways, which in turn can contribute to its mind-enhancing effects. At certain doses, some cannabis varieties for instance have been reported to have anxiolytic (anxiety-reducing) and euphoric effects during a high.

Think about how these effects may incline users to open up to others and to themselves, to think about death and other difficult subjects without fear. In the face of deep and troublesome aspects of their lives, users may feel that they are better able to face their problems and see them from a different angle. Many users have said that the high allowed them a better understanding of subtle humor.

It is crucially important to the research of the cannabis high that we listen to users – and pay close attention to detail when they are describing their experiences with these mind enhancements. These reports not only give us a good starting point for investigating the temporary cognitive effects of the high, but they also show us how important and meaningful these experiences can be. Let me quote here only a few examples of some astounding reports.

Beth Amberg, a contributor to Lester Grinspoon's magnificent website collection *marijuana uses.com*, is one of those many who reported how marijuana helps her to better remember past events:

> *"Perceptions are heightened tonight, my mind unencumbered and slippery. I'm still so close to the wonder and sensations of the past. My thoughts are swimmy-silvery fountains of assorted memories, the novelty generator of marijuana turning its freshness backwards into history. My past selves have awoken: their experiences aren't distant; they happen again as I read and remember. The shimmering glaze on memory has opened up and let me back in for the night."*[15]

The saxophone player Milton "Mezz" Mezzrow worked with other jazz musicians like Bessie Smith, Louis Armstrong, and Bix Beiderbecke. His colleagues highly esteemed not only his musical talent but also the quality of his marijuana. Mezzrow wrote about the influence of a cannabis high on his playing:

> *"The first thing I noticed was that I began to hear my saxophone as though it were inside my head . . . Then I began to feel the vibrations of the reed much more pronounced against my lip . . . I found I was slurring much better and putting just the right feeling into my phrases . . . All the notes came easing out of my horn as they'd already been made up, greased and stuffed into the bell, so all I had to do was blow a little and send them on their way, one right after the other, never missing, never behind time, all without an ounce of effort . . . I felt I could go on playing for years without running out of ideas and energy. There wasn't any struggle; it was all made to order and suddenly there wasn't a sour note or a discord in the world that could bother me . . . I began to preach my millenniums on my horn, leading all the sinners to glory."*[16]

For thousands of years, marijuana and hashish users have told us about the aphrodisiac potential of cannabis. In Erich Goode's book "The Marijuana Smokers" (1971), an Atlanta woman reports:

"The most terrific experiences I've had while stoned have been sexual encounters. I finally learned how sensual my body is, and I can say without a doubt that marijuana contributed to this discovery. I often get high before making love. My body responds in a more fluid, warm manner, with visual imagery intensified, and every touch sending notes of ecstasy to my brain. No, I have not become a "loose woman" because I smoke pot. But I'm a lot looser than I was ten years ago. I'm not sure how much of this is due to grass, and how much is because of my personal growth; for me, the two go together and can't always be separated. But I do know that my sexual expression has been greatly enhanced since I started getting high." [17]

These are only three quotes from hundreds of detailed reports about cannabis enhancements from skilled and experienced users which I analyzed for my research.

On the other hand, we also have many other reports of people who found that a cannabis high negatively interfered with many of their activities, for example, reports from users who found that the cannabis high confused them or made them too tired or introspective for sex or other activities. But it should be clear now that it does not make sense to simply count these reports as evidence that cannabis does not have the positive potential for these activities. When looking at the role of the elevated state of a high for creative purposes, for pattern recognition, introspection, or other activities, many analysts take a superficial look at anecdotal reports from users to find that it is a fifty-fifty situation – with half of the users saying they experienced various enhancements, whereas the other half reports that the high made things worse for them in various ways. Commentators often tend to conclude that therefore, it must be a myth that cannabis has the potential to enhance the abilities in question. The surfboard metaphor should make it clear that this evaluation approach is highly questionable. Many negative reports come from consumers who have used cannabis in the wrong way, taking a dosage too high to handle in a certain situation, using bad quality cannabis from a black market, or consuming cannabis in the wrong social context without skills and knowledge. Also, many of those consumers are negatively biased.

If we want to evaluate and research the positive potential of cannabis, it will not be enough to make statistics by superficially counting positive and negative reports about cannabis use for various purposes. If we want to study the positive potential of surfboards, it will not help us to ask thousands of novices to get on a surfboard and go out with it in the ocean. Naturally, many of them will come back with negative reports. Even if we scientifically set up to study with them and if we make a very precise count of their negative and positive reports, the outcome will tell us nothing about the real potential of surfboards. As trivial as this may sound for the case of surfboards, many do not get this point when it comes to the evaluation of the positive potential of cannabis.

Cannabis as a Tool: Lessons to be Learned

One lesson of the surfboard metaphor, then, is as simple as it is important: if we want to learn about the positive potential of cannabis, we will have to listen to the reports of skilled, knowledgeable users who consumed good quality cannabis under favorable conditions. And of course, this will only be the start. What we need is a critical assessment of these stories and reports about cannabis enhancements. From there, we can come to build interesting hypotheses about the psychoactive effects of cannabis and how it may be used by skilled and knowledgeable users for various mind enhancements under the right condition. In the essays in this book, I will not only do this, but I will also integrate knowledge from many other scientific disciplines to give those hypotheses further credibility.

There is also a personal lesson to be learned from the surfboard metaphor: Whatever reasons you may have for using cannabis, if you decide to do so, I would recommend you read about cannabis beforehand and learn from skilled users who know what they are doing. Whatever reasons you may have to go and ride ocean waves on a surfboard if you decide to go out there, would you not want to learn from a knowledgeable and experienced teacher?

The book you are holding in your hands hopefully helps to come to a deeper understanding of the mind-enhancing potential of cannabis. For practical guidance, I have written a minimalist hands-on guide for those who have decided to use cannabis for inspirational purposes *"The Art of the High: Your Guide to Using Cannabis for an Outstanding Life."*[18]

Whether you are riding 'high' waves of your mind or ocean waves on a surfboard, skills, and knowledge will get you to a point where you will 'step into liquid' and understand what legendary surfer Bill Hamilton meant when he once said:

"Surfing equates to living in the very moment of 'now'. When you ride a wave you leave behind all things important and unimportant, and the purity of the moment is upon you."

(5) A cannabis plant still in the germination phase.

II. Baudelaire, Ramachandran, and Colors Containing Music

Charles Baudelaire and the Club des Hashashins

The *Club des Hashashins* was founded in 1844 in Paris, only four decades after Napoleon's troops had retreated from their catastrophically failed occupation of Egypt, bringing hashish as their discovery with them. Napoleon's troops had first encountered the use of hashish through local Muslim dealers and kept consuming it during the occupation despite harsh prohibitionist orders imposed by the French government. Within a few decades after their retreat, the use of hashish for recreational as well as medicinal purposes had found its way into the societies of France and other European countries.

Many of the members of the *Club des Hashashins* belonged to the French literary and intellectual elite of their time: founding members psychiatrist Jacques-Joseph Moreau and writer Théophile Gaultier, poet and writer Charles Baudelaire ("The Flowers of Evil"), writer Gérard de Nerval, the painter Eugène Delacroix, as well as writers Alexandre Dumas ("The Three Musketeers") and Gustave Flaubert ("Madame Bovary"), to name only a few. Their goal was to experiment and explore the hashish experience for various intellectual and artistic endeavors – and they did so by ingesting large doses of hashish concoctions. Interestingly, Baudelaire made many important observations concerning the cannabis high, but his conclusion concerning hashish as an enhancer for creative purposes was rather critical, claiming that it would only "magnify" impressions and thoughts of a user which are already there. In another passage he seemed to be more positive, but also warned:

> ". . . that hashish gives, or at least increases, genius; they forget that it is in the nature of hashish to diminish the will and that thus it gives with one hand what it withdraws with the other; that is to say, imagination without the faculty of profiting by it. "[19]

Little did he know that he was maybe the first one to describe one of the key psychoactive effects of cannabis responsible for the creative enhancement during a high when he observed that *"sounds take on colors and colors contain music. "*[20] While his description of this altered state of mind is fascinating, it sounds just as exotic as hard to believe. However, we will see that only recent findings in the neurosciences have revealed the validity and significance of Baudelaire's observation for the understanding of the cannabis experience.

The Phenomenon of Synesthetic Experience

Baudelaire's statement about the perception of sounds taking on colors during a high describes what we would now call a *synesthetic* experience. The expression *synesthesia* stems from the Greek words *syn* ("together") and *aesthetics* ("sensation"). Today, synesthesia is known as a neurologically based phenomenon in which stimulation of one sensory pathway automatically results in correlated experiences in a second sensory pathway. In a synesthetic experience, for example, during an LSD trip, you might have a corresponding metallic taste experience associated with the tactile experience of touching a metal object, or you may see correspondingly dancing colors when listening to a Jimi Hendrix guitar solo – an effect simulated very effectively today by music visualizing software. Synesthetic experiences can be induced by psychoactive substances like cannabis, psilocybin mushrooms, or LSD, but we also know that there are also natural-born *synesthetes* experiencing various forms of *synesthesia* in their lives.

The phenomenon of synesthesia had not been described before in the scientific literature when Baudelaire experienced this phenomenon, so Baudelaire invented a name for it: "equivocations" ("*équivoques*").[21] The first detailed description of *synesthesia* would only come a few decades later – in 1883 – from the Victorian multi-talented genius Francis Galton, a half-cousin of Charles Darwin.[22] For a long time since then, synesthesia had been largely ignored – and only in the last decades did neuroscientists rediscover this phenomenon to find how crucially important it may be for human cognition and creativity.

Synesthetic experiences have been described by many users of cannabis usually only after consumption of relatively large doses. It is not surprising, then, that not only the hashish-eating Baudelaire but also the famous cannabis user and American writer Fitz Hugh Ludlow would describe this effect; both of them experimented with copious amounts of ingested hashish concoctions or hash oil. In 1857, Ludlow wrote about his synesthetic experiences: *"Thus the hasheesh eater knows what it is to be burned by salt fire, to smell colors, to see sounds, and, much more frequently, to see feelings."* [23]

Baudelaire and Ludlow are not the only ones who described synesthetic experiences during a strong marijuana high. For his study "On being Stoned" (1971), Harvard psychologist Charles Tart sent out questionnaires to hundreds of students to ask them about their high experiences and found that the phenomenon of synesthesia is a commonly reported effect of cannabis for high doses.[24]

Now, while it seems interesting to see vexing colors during a high when listening to music, or to "taste colors," why should this explain the often reported enhancement of creativity during a cannabis high? What is the connection? Also, if this effect only occurs under higher doses of marijuana, could it be relevant to creative abilities when it comes to the usual that high modern consumers usually experience? If we want to understand how the synesthetic effect can play a role in the enhancement of creative thinking, we have to take a look at the groundbreaking work of Vilayanur Ramachandran and other neuroscientists who changed our understanding of the phenomenon.

Synesthetic experiences are not always triggered by psychoactive substances only. Naturally *synesthetic* people experience various forms of *synesthesia* because of a genetic predisposition. Famous natural *synesthetes* include composers Nicolai Rimsky-Korsakov and Leonard Bernstein, jazz legends Duke Ellington, the drummer Elvin Jones, the painter David Hockney, and the writer Vladimir Nabokov, a "color-to-grapheme"- *synesthete*. A *synesthetic* person with this often-occurring variety of *synesthesia* experiences black letters as colored. For example, a *synesthete* might experience the letter "a" as red and the letter "g" as green, and this will consistently be so for this individual throughout his lifetime.

A synesthetic person with grapheme-to-color *synesthesia* sees words and numbers colored as in the following picture, even if they are printed in black.

SYNESTHESIA
0123456789

As opposed to Nabokov's genetically triggered *synesthesia*, *synesthetic* experiences caused by psychedelic substances, or by a stroke, are called "*adventitious synesthesia.*" As observed above, the effects of adventitious synesthesia are more common for "trips" under the influence of psychedelic substances like LSD or psilocybin (the active ingredient in magic mushrooms) but are also frequently reported by cannabis users for higher doses.

In their highly influential article "Hearing Colors, Tasting Shapes" published in Scientific American in 2003, neuroscientists Vilayanur Ramachandran and Edward Hubbard state:

> "*When we began our research on synesthesia, we had no inkling of where it would take us. Little did we suspect that this eerie phenomenon, long regarded as a mere curiosity, might offer a window into the nature of thought.* "[25]

According to Ramachandran and Hubbard, natural synesthesia is caused by the cross-activation of neighboring brain regions. Cross-activation concerns the chemical exchange of neurotransmitters, which travel between various brain regions. Usually, neighboring brain regions like the "hearing center" in the temporal lobes and the color signal-procession area produce inhibiting neurotransmitters to minimize cross-activation of those regions. Our brain manages to actively inhibit chemical interchange between these signal-processing areas and, thus, helps to keep our senses separate.

Ramachandran and Hubbard assume that synesthetic experiences occur when these biochemical inhibitors are blocked, thus allowing two neighboring brain regions such as the hearing center and the brain area that receives color signals to exchange signals. As a result, on the subjective level, this leads to cross-sensory channel experiences – thus, an auditory experience of a Shostakovich symphony may stimulate the visual area and lead to the experience of an accompanying firework of colors.

It is well known that psychedelic drugs such as LSD or cannabis can produce synesthetic experiences in otherwise normal people, but the neurophysiological consequences of this process do not seem to be explored much yet. Still, based on the discoveries of Ramachandran and Hubbard, it makes sense to claim that these substances have a short-term effect on those inhibiting neurotransmitters and allow for the unusual cross-experiential phenomenon of synesthesia.

The Relevance of Synesthesia for the Understanding of Human Consciousness

"Fine," you may say now, "but again, what is the freak show good for? It certainly makes for a beautiful and unusual firework of the senses, but can't we get that easier if we go and watch music videos, where we find music transposed into images? Apart from being entertaining for a while, is it not just useless and disturbing to have those sensual crossovers? It may be fascinating to smell a movie, but can we learn anything here?"

The answer is an emphatic "yes". I am convinced that the effect of synesthesia could be one of the central effects of the cannabis high. As Ramachandran and

(6) A Northern Lights #5 Haze cannabis variety still in the growth phase about two weeks after its germination.

Hubbard observe, natural synesthesia is seven times more common in creative people than in others. They suspect that this may have to do with the enhanced ability of the *synesthete* to set up links between seemingly unrelated domains, a capacity that importantly subserves the ability for inventing metaphors, when we for instance say that "the pigeons *fountained* into the air." Remarkably, many of our everyday metaphors are synesthetic in nature: common synesthetic metaphors would be 'loud colors, dark sounds or sweet smells', but of course, they can find much more sophisticated expressions in literature and poetry.

Ramachandran and Hubbard point out that inventing metaphors and linking seemingly unrelated concepts and ideas are an integral part of creativity. They also suggest that enhanced creativity may explain the advantageousness of the *synesthetic* trait, which in turn would explain the survival of the genetic predisposition in some individuals from an evolutionary perspective.

The synesthesia specialist Richard Cytowic, as well as Ramachandran and Hubbard, believe that in essence, all humans are "closet *synesthetes*". Cytowic suggests that our perception may be inherently "holistic", but that after a pre-conscious information process, only the distinct sensual information enters our consciousness. Ramachandran and Hubbard point out that *synesthetic* effects can be observed in "normal" (and non-drugged) humans as well. They use an example from the famous German *Gestalt* psychologist Wolfgang Köhler to illustrate their hypothesis:

When normal people have to decide between "Bouba" and "Kiki" as names for these two graphics, 98% judge the left one to be Kiki and the right one to be Bouba.

If you look at these two inkblots, which one would you call 'Bouba' and which one 'Kiki'? I hope you have made the right choice. If not, there may be something seriously wrong with your angular gyros region. Ninety-eight percent of probands would call the left-hand blot 'Kiki' and the other 'Bouba'. Ramachandran and Hubbard believe that the angular gyros region is responsible for 'extracting' the abstract common denominators between the graphic shape of the blobs and the names, like the gentle undulation of the sound of the name Bouba and the gentle curves of the inkblot. Further, they note that the *angular gyros* is usually considered the brain region where sensorial information from touch, vision, and hearing flow together and is involved in the synesthetic condition. They further believe that the natural synesthetic trait may bring with it not only excess communication between the sensual information centers but could also help to 'link' seemingly unrelated concepts and ideas:

> *"It has often been suggested that concepts are represented in brain maps in the same way that percepts (like colors or faces) are. One such example is the concept of a number, a fairly abstract concept, yet we know that specific brain regions (the fusiform and the angular) are involved. Perhaps other concepts are also represented in non-topographic maps in the brain. If so, we can think of metaphors as involving cross-activation of conceptual maps in a manner analogous to cross-activation of perceptual maps in synesthesia. "*[26]

Cannabis and Synesthesia

If we take this assumption seriously, then we might also have the beginning of a possible neurological explanation of an enhancement of creative thinking under the influence of cannabis. During a decent cannabis high, users may not experience full-blown synesthetic experiences such as seeing colors corresponding to a guitar solo. Subconsciously, however, a *pre-synesthetic* effect may allow for facilitated communication between brain areas which represent structurally similar (or '*isomorphic*') percepts or notions. In other words, during a high, we can more easily find associations between structurally similar images, concepts, ideas, or patterns, like the linking of the visual curved form of the "blob" and the auditory representation of the word "Bouba".

This observation is fundamental. This kind of excess communication does not only allow us to create metaphors but generally enhances our ability for what I would like to call "*isomorphism extraction*". "Isomorphy" is a technical term for similarity. Less technically speaking, isomorphism extraction is the abstraction process that allows us to find a similarity, or a 'common denominator', between various patterns, structures, forms, events, and concepts. This could be an abstract similarity between our representation of the sound of "Bouba" and the visual representation of the actual 'blob'; which could be subsumed as "soft and round". But it could also be a much more complex similarity. For instance, you may perceive a similarity between the feeling depicted in the song Flamenco Sketches of the recording of Miles Davies's *Kind of Blue* to the mood you once had sitting on an early summer evening in New York City's central park on a bank facing a little lake, and reflecting on your life.

We have many detailed reports from cannabis users which report how their ability to recognize new patterns was enhanced under the influence of cannabis. We usually think of visual patterns when talking about pattern recognition. Yet, of course, patterns are everywhere, and pattern recognition is one of the most fundamental of human abilities: we can smell the olfactory pattern of something burnt, we recognize a certain pattern in the taste of a great Sauvignon wine, recognize a game pattern in the way a friend plays chess, or we recognize the inimitable sound and style of John Coltrane playing his saxophone in various recordings.

Many cannabis users have reported how they have found new patterns during a high. Here is the curious report of an anonymous contributor "Rob" to Lester Grinspoon's website "marijuana-uses.com," who writes in a letter to his parents about how he once became aware of a pattern of "timid rigidity" in his walking style:

"I was also fortunate in that I got high that night, as many reefer virgins report not getting high the first time they smoke. An amazing understanding came to me while walking home. As I strolled along the tree-lined sidewalk carrying on a conversation with a friend, I felt an awkward stiffness in my stride, realizing with each step a timid rigidity. I have since altered my manner of walking to a confident, open gait of long strides and silent footsteps." [27]

In another report, a woman named "Martha" got high and suddenly recognized a pattern in her friend's behavior and her character:

"During the conversation with Alice and Karl, I realized that she was being very self-conscious, and kept stepping back out of herself. I looked at her and thought, "That's a whole new way of looking at Alice." I had never seen her insecurities so palpable before. (...) I suddenly understood that her insecurity was a key to her personality (...)" [28]

If Ramachandran and Hubbard are right, then this enhanced pattern recognition during a high might at least to some degree be due to a *pre-synesthetic effect*. I find it interesting to note in this context that both Rob, as well as Martha, would use remarkable synesthetic metaphors in their descriptions. While Rob names his walking style 'timid', Martha says that she has "never seen her *insecurities* so *palpable* before." The use of those metaphors here may well be a coincidence, but I think that in the light of this essay and its implications, we can see how it may not be so coincidental after all.

Let me sum up. We can see how cannabis might even at lower doses have a *subliminal synesthetic* effect, leading to a better ability to make associations between seemingly unrelated concepts or structures, invent metaphors, and recognize patterns. Recent neuroscientific research on *synesthesia*, then, may give us the beginning of an explanation of the many reports of artists, musicians, writers, scientists, and comedians who reported not only various creative enhancements during a high but also described an enhanced ability for pattern recognition.

Baudelaire, then, probably understated the potential of cannabis to enhance our creativity. Cannabis does not only "magnify" one's thoughts and sensations, but it causes an interesting process in our brains that might help us in much more interesting ways to come up with new ideas and perceptions.

Surprisingly, however, when we go back and take a closer look at Baudelaire's statement about "equivocations", we find that he not only clearly described the now well-known synesthetic cross-modal *sensory* effect, but he also seemed to have intuitively grasped the intellectual creativity-enhancing effect described by Ramachandran and Hubbard himself, namely, that:

> *"The most singular equivocations,* **the most inexplicable transpositions of ideas take place.** *Sounds have a color, colors have music. Musical notes are numbers ... "*[29]

In a way, then, the great poet Baudelaire not only left

us one of the first descriptions of synesthetic experience, but he also hinted at how this phenomenon could be involved in the enhancement of creative thinking during a cannabis high.

As we will see in the next chapter, there are many more ways in which a cannabis high could interestingly enhance creative activities – but, also, ways in which it could have a negative impact on them. If we want to find out more about how a high could help us to be creative, we will have to first come to a better understanding of creativity.

(7) A Swazi Safari variety of cannabis after about two weeks of growth. The two rounded leaves on the bottom of the plant are the cotyledons or cotyledons, which do not yet have a specific pattern like the leaves which grow later.

III. Cannabis, Creativity, and Cognitive Liberty

"There is no doubt that creativity is the most important human resource of all."
Edward de Bono, psychologist, 1933-2021

"You are the task."
Franz Kafka, writer, 1883-1924

Does cannabis enhance creativity? Countless columnists and scientists have asked and discussed this question in the last years – usually coming to negative or at least contradicting results.[30] Many cannabis users have asserted that cannabis has indeed helped them with their creative work, whereas many others reported that the cannabis high seriously negatively interfered with their creative output.

Creativity is one of the main cognitive resources for humans to deal with and overcome the challenges of life. Creative ideas and behaviors cannot only be used to produce art or music or to come up with new scientific findings. They can help us to find a partner for life, to find an unusual way to cope with the stress inflicted on us by co-workers in an abusive workplace, come up with a new strategy to generate income, or personally grow. In the last decades, advances in cognitive sciences and psychology have made it increasingly obvious that creativity is one of the main positive driving forces for individuals as well as for important developments in human societies. Creativity shapes our society not only through the inspiring work of writers, artists, or musicians, but also through innovations in the sciences, technology, the business world, the financial sector, architecture, and countless other areas of life.

As hard as it may be to define and describe, creativity is no doubt one of our most important human resources, especially as humanity faces existentially-threatening challenges like climate change, the rapid and powerful rise of AI, the rise of autocrats and dictators in so many countries, the negative impact of modern social media on our democracies, or the threat of deadly nuclear weapons in the arsenals of many countries in the world.

As more and more countries have liberalized their laws concerning the medical and general adult use of cannabis, an increasing number of users have gained legal access to high-quality cannabis as well as better information about its use, its benefits, and risks. Cannabis users today are now better equipped than ever to explore the mind-altering potential of this astounding plant. In the last years, we have seen an avalanche of reports, documentaries, studies, and articles about cannabis and its effects, as well as about various phytocannabinoids used in medicine, and the endocannabinoid system.

So, do we now have a better answer to the question of whether cannabis enhances creativity? What's the current verdict?

My verdict is, first, that there is something seriously wrong with our question. In this essay, I'll explain why I think we need to find better questions and will suggest some preliminary answers to them based on our current knowledge. Also, I'll make suggestions for future routes of scientific investigations and give some practical advice for cannabis users on how to better use cannabis for practical purposes.

The Art of Questioning

"We get wise by asking questions, and even if these are not answered, we get wise, for a well-packed question carries its answer on its back as a snail carries its shell."

James Stephens, Irish poet, 1880-1950

"Wait," you say, "how can a question be wrong? Is the question 'Does cannabis enhance creativity?' not a perfectly reasonable and innocent question to ask?"

Well, frankly, no. Not after all we already know. We should be able to ask better questions based on what we know about creativity and the effects of cannabis today.

Questions are tools that help us to learn and gain knowledge; they can be "bad" and misleading in many ways, and some of them are even designed to be misleading. Assume I ask you in a meeting with some of your friends: "why didn't you help to prevent the bankruptcy of your father's ice-cream business?" Assume, also, that your father owns an ice-cream business, but it is doing just fine. My question is wrong, then, in the sense that it presupposes a situation that is not true. Questions can carry presuppositions, and these presuppositions can be false or misleading.

As we will see, "Does cannabis enhance creativity" carries various highly problematic presuppositions. It is, additionally, too ambiguous and vague to help us to gain significant knowledge. Let us take a look at some of the problems with this initial question so that we can come to formulate more interesting questions – and then suggest some answers to those.

First, the question "Does cannabis enhance creativity" presupposes – or at least strongly suggests – that the real issue is whether cannabis under all circumstances and for everybody, no matter what dosage, and no matter in which situation, enhances creativity. It either does enhance creativity, or it doesn't. It is obvious, however, that cannabis can only enhance creativity given a whole list of other favorable conditions, for example a specific dosing range, or a specific context/setting in which somebody uses cannabis. A user relatively new to cannabis will probably not be so great at a complicated creative verbal task administered in a test after smoking three joints packed with a new cannabis variety containing almost 30% THC (aptly called "99 problems").

The more interesting question, then, would be:

"Can cannabis enhance creativity?"

Again, this first modification of the question sounds trivial. Yet, I don't think it is trivial at all if you look at studies and reports about the topic of cannabis and creativity. Often they highlight the fact that high doses of cannabis impair your creative output and do not take into account that other factors such as the (mind)set and the setting (context/environment, other people present), which are substantial factors for the effects of any psychoactive substance on the mind.

If we ask "Can cannabis enhance creativity?" it becomes obvious that we need to answer more questions: Which are the factors that can affect or modulate the effects of cannabis on our mind and our creative abilities? How much and how exactly do they contribute to the effect of cannabis in this aspect? We will come back to these questions later.

Now, let us address an important ambiguity in the question. Usually, when somebody asks if cannabis can enhance creativity he inquires about the short-term effects of the cannabis high. Some critics however, also seemed to confuse this with questions about the long-term effects of continuing cannabis use, or maybe even the question of how daily use of high doses of cannabis affects one's creative career. But these are entirely different questions: Maybe cannabis has a great potential to temporarily enhance your ability to generate some great ideas for a novel, but it might still have a negative impact on your writing career if you use it daily in high doses.

In this essay, I will focus on the short-term effects of the cannabis high and will not go into the broader issue of how long-term use may affect your creative career. First things first, I'd say. Once we better understand how the short-term effects of cannabis can affect our creative abilities, we can then set out to take a more informed look at the long-term effects.

Our new question, then, would be:

"Can a cannabis high temporarily enhance creativity?"

We are not done yet with refining our question, though: the new version of the question still has some major flaws and carries some misleading presuppositions. If we want to find out why, we first need to take a closer look at the concept of "creativity" and come to a better understanding of the phenomenon.

Creativity: A Minimal Characterization

Many who have delved into the subject of cannabis and creativity have made it seem like we would have one well-defined and uniform process or activity we call "creativity". Is this so? What is creativity?

We can find dozens, maybe hundreds of diverse definitions and characterizations of creativity in literature and science. So let us first take a look at a possible *minimal definition of creativity* on which many scientists in the field could agree, and take it from there. In the last decades, most scientific definitions of creativity can be summarized by a minimal characterization given by psychologist and creativity expert Robert Sternberg: Creativity (a) leads to the production of something *novel or original*, and (b) the product is something *valuable or useful*.[31]

This definition is of course vague to a certain degree: it is not clear what we would regard as entirely novel, and it also does not say much about the standard which should be applied to whether something is valuable or useful. More importantly, though, there seems to be something essentially missing in Sternberg's definition. If an earthquake shakes up rocks that roll down to accidentally form into a new and funky looking shape of a rock pile with a hollow core that can be used by humans as a house or shelter, then can we legitimately say the earthquake has *"created"* something *novel* and *original*, and something very *valuable* and *useful?* Most of us would be somewhat reluctant to call this a *creative* process, however. Although we may choose to call nature "creative", metaphorically speaking, we usually think of creativity as a process that happens in higher evolved lifeforms, organisms, or systems.

Hundreds of Different Creative Activities

The observation about the rock pile points to an important aspect: the minimal definition of creativity given by Sternberg does not give any characteristics of the *mental processes or abilities* involved in the creative process. We usually think of creativity as a highly refined human skill or a skill possessed only in a few highly developed animal species like chimps, or maybe in current or in a few current or future AI systems. From a cognitive science perspective, the processes involved in creative production are not only highly complex. Hundreds of different creative processes involve the concertation of a whole variety of different mental and physical skills. For illustration, let's look at two examples of human creative activities:

(a) Inventing a basketball move

(b) Writing a poem about a street scene in Rome

Let's assume you are a basketball player in the last 5 seconds of a game, it's still a tie, you are dribbling and close to the opponent's basket. You know your defender knows you inside out. You need to come up with a creative and new move, fast, a move that will trick even him. There are many mental skills involved: you need to empathically understand your opponent and "read" his movements as you are dribbling, you need to keep attending to his moves while you come up with your new idea. You have to keep moving without losing the ball, and you need to keep track of the merciless clock ticking down the last seconds of the game. Simultaneously, you have to get your great idea at the right moment. You remember a moment from your adolescence when a Chinese Tai Chi grandmaster asked you to hold on real tight to his lower arm with both of your hands. He announced that he could easily remove his arm. You waited, holding your breath in tension of the anticipated move.[32] A few seconds later, he confused you for a split second by gently slapping the back of your clutching hand. Back then, you were surprised by this seemingly nonsensical move and lost your muscle tension for a split second when he freed his arm by a gentle, quick move. Let's assume you now use this memory and transpose it to your situation to invent a new fake, designed for this situation in the basketball game. You pretend to make a pass with one hand, with your hand formed like a fist. For a split second, your opponent is puzzled, just as you were back then. You jump in the air and shoot. You gained exactly the time you needed to shoot the ball, undisturbed. Like your reaction many years ago, your opponent's confusion left him dysfunctional for a short time. For your basketball fake to be useful, however, you have to execute that idea flawlessly, control your body perfectly in only a few seconds, and, then, redirect your attention quickly to focus on the basket and perfectly execute your shot.

Let's assume for a moment that you are high and under the influence of cannabis, and that because of this, you can indeed retrieve distant memories better, and that it also enables you to come up with a great idea, an adaptation of the Tai Chi fake move you have previously seen. Will a cannabis high be helpful in this situation? Not necessarily, if we evaluate the creative activity and its value of it in the context of the game. If for some reason the high helps you to generate this idea but also interferes with your sense of orientation or your sense of time, you may have a great and novel idea, but it will not lead to a valuable and useful outcome for your creative activity. As you come up with a new fake move, your distorted sense of time maybe makes you wait for too long. Or you lose the ball during the fake to your opponent because you are too focused on your thinking and miss his movement towards the ball. Maybe the high helped to focus your inner attention on your memories, but now your lack of attention for your opponent spoils it all. There are many ways in which a high could have a negative impact on your playing despite helping you to generate an idea.

Now, consider a different creative activity: writing a poem about a street scene in Rome. Assume you want to write a poem describing a certain perceived mood of an evening street scene at the famous Spanish Steps at the *Piazza di Spagna* in Rome. You have time – maybe hours, or days. You perceive various scents, sounds, voices, moods of other guests, music on the street, and then slowly come to merge all these impressions into a composition of words based on your thoughts about the life around you, the world, life. You may need to retrieve remote verbal associations, come up with metaphors, and maybe make your words rhyme in a certain style that you retrieve from memory. The creative process could start while you sit there taking notes and stretch for days when you remember certain aspects of the scene and start forming your first notes into a more structured text.

For the writing of a poem, a cannabis high could still be helpful even if it distorts your sense of time, or even if it for instance interferes with your hand-eye motor control. Your handwriting may look a bit different, but that would not lessen the quality of your poem.

Those two examples of creative activities illustrate two important points for us:

First, if we want to know how cannabis, or any other

psychoactive substance, affects creativity we need to look at various *creative activities* which involve the concertation of various cognitive and motor control skills. We have to be prepared for the fact that a cannabis high may affect these creative activities in completely different ways. And that, again, means that we have to refine our question. Now we have to ask:

Can a cannabis high temporarily enhance *certain creative activities*?

Second, if we take a closer look at the example of the basketball player, we need to talk about what we could call "cancellation"-effects during a high. For some activities, various cognitive effects of cannabis may be beneficial, such as enhanced memory retrieval and the ability to transpose far-fetched patterns to different situations, as the basketball player does use an analog to the Tai Chi master's trick for a basketball fake. But other effects of the high such as a distortion of time perception could cancel out the beneficial effects in the sense that the resulting action isn't *useful* or *valuable* as a whole anymore. Another example would be a jazz saxophonist who feels that he generates great ideas during a high, but cannot play along with his band anymore because he loses his sense of time too much. These "cancellation" effects are intuitively very easy to understand, but highly underrated so far in the literature on cannabis and creativity. As we will see, the existence of cancellation effects plays a big role in constraints on how to construct studies on cannabis and creativity.

These cancellation effects during a cannabis high exemplify a general aspect of psychoactive substances. All of those substances bring certain temporary cognitive impairments, too. Typical cognitive enhancers like amphetamines or cocaine may make you more awake, alert, and speed up aspects of your thinking, but you also end up losing some of your ability to empathically perceive and understand others around you. It depends on our goals and needs if we consider the general effect on our cognition by those substances to be beneficial in some aspects.[33] Again, we are reminded of the wise words of Baudelaire when he wrote about hashish: *"thus it gives with one hand what it withdraws with the other."*

(8) The variety Skunk Auto after about a month of growth. The addition of "Auto" in the name indicates that breeders used genetics of Ruderalis cannabis plants to make their new hybrid plants begin to flower automatically, regardless of the light cycle. Normally, cannabis plants only start to flower when the light time during the day decreases like in autumn - a disadvantage for indoor growers who have to simulate two different light cycles with lamps.

The Creative Process: Four Phases

So, if we talk about creativity we need to understand that various creative activities involve the orchestration of very different mental skills and physical abilities. And it gets even more interesting if we look closer at various phases of creative thinking.

Almost a century ago, social psychologist Graham Wallas drew on observations made by the German physicist Hermann von Helmholtz and the French mathematician and physicist Henri Poincaré to describe a model for the creative process that is still widely considered to be useful today.[34] Wallas distinguishes four phases:

(1) Preparation

You collect information about a certain problem or task, maybe discuss it with others, etc. this may take from a few minutes to years or decades.

(2) Incubation

You are not consciously thinking about your creative project or problem. Wallas points out that there may be various unconscious thought processes going on that work towards a solution. He also stresses that mental relaxation will help in this phase to move towards phase 3, illumination.

(3) Illumination

The Aha-Moment and some unspecified time that leads up to it. Suddenly, seemingly out of nowhere, the solution comes in a flash. Usually, we cannot tell how we came up with this and Wallas stresses that the thought processes leading up to this moment are either subconscious or only in the "fringe of our consciousness".

(4) Verification

According to Wallas, this phase is more like the illumination phase, which means that the thinking is more consciously directed. Here we edit our idea, evaluate it critically, and refine it.

Importantly, Wallas himself remarks:

"In the daily stream of thought, these four different stages constantly overlap each other as we explore different problems."[35]

I would add that some creative projects like writing a novel may demand several hundreds of little creative solutions. What is my headline for this chapter? How do I describe the character of my main proponent? What name should I give her? What is a metaphor that summarizes the theme of my chapter? How can I visualize the conflict between the two families involved in a conflict?

If we want to take into account that creative activities come in various phases, then, we have to refine our question again:

"Can a cannabis high temporarily enhance a creative activity in a certain phase?"

Look at an artist during the process of painting, for instance. A cannabis high may help her to enhance her imagination and, thus, may lead to an unusual, novel mental image. Maybe the high helps her to make wider associations and to lead to an unusual idea for painting a portrait with unrealistic colors expressing the emotions of the portrayed rather than the perceived colors of the skin. In a different phase of painting, however, the high may have a negative impact. A painter may for instance feel that if she wants to execute her idea and paint, a high may not be helpful for her hand-eye coordination while drawing, or she feels too distracted to get into details because of her racing mind.

Simultaneous and Consecutive Cancellation Effects

We can now also see that there could be different types of cancellation effects when it comes to cannabis and creativity: simultaneous cancellation effects, and consecutive cancellation effects. In the example of the basketball player, I have described a simultaneous cancellation effect: while he generates a great "highdea" for a fake, his motor skills, attention, and sense of time may get disturbed by other effects of the high. The total effect of the cannabis high, then, could be detrimental to his creative activity.[36]

Here is an example of a consecutive cancellation effect: assume a painter has an innovative idea during a strong high because of an enhanced ability to imagine novel sceneries. He feels too "stoned" to paint, so he rides off his high, and his jumpy and quick associative stream of thoughts during the high takes him to all kinds of places. One hour later, though, when he sits down to paint, the idea he had before could be long forgotten. Again, some of the effects of the cannabis high cancel the positive influence and could lead to a detrimental overall outcome of the high on the creative process as a whole.

It is easy to see that cannabis users could avoid cancellation effects with some knowledge and some acquired skills. As I have described in my book *The Art of the High: Your Guide to Using Cannabis for an Outstanding Life*[37], they should learn about dosing, the different effects of various cannabis varieties, the impact of set and setting, and various other factors that can help them to avoid cancellation effects and to get the best out of their high not only for creative purposes but also to experience other useful mind enhancements.

Substantial, Essential, and Characteristic Enhancements

Here is another aspect of our question we have to address: What *kind of enhancement* do we mean if we ask if cannabis can enhance creative activities? We can distinguish at least three interesting categories of possible enhancements:

1) substantial enhancements

2) essential enhancements

3) characteristic enhancements

A substantial contribution of the cannabis high to creativity would change one or several mental or physiological processes in a way that leads to a *substantially* better creative result, qualitatively speaking. Let's assume – and the evidence so far is pretty strong for this claim – that some cannabis varieties can help us to relieve stress and to relax, to turn our attention away from our daily sorrows and woes. Modern science on creative insights produced some good evidence that relaxation during some phases of the creative process may indeed be very helpful for the generation of novel ideas.[38]

In certain stages of the creative process, the mental relaxation and physiological muscle relaxation resulting from a cannabis high may then *substantially* help you to come to important creative breakthroughs. Yet, even if this contribution of the high may be substantial – is it an essential contribution to the creative process? A meditation, closing your eyes, or breathing could all lead to relaxation and help the creative process, but would these activities contribute something that we consider essential, something that specifically enhances the quality of the creative process?

Here is another example of an effect that we would probably consider to be substantial, but not essential: Substances like *methylphenidate*, the active ingredient in *Ritalin*, or cannabis may help you to sustain your focus of attention longer on certain tasks. So, they could help many creative activities substantially especially in the preparation phase and in the verification process. Arguably, though, most of us would not consider this stronger focus as essential to the enhancement of creativity, as compared to, for example, in the case of an enhancement of our ability to imagine non-existent objects or to recognize new patterns.

Similarly, cannabis can be mood modulating (causing a profound state of *euphoria or bliss*, for example), and can affect our ability to improve our creative output. There is scientific evidence showing that our mood indeed has a strong and maybe sometimes substantial effect on our creative output.[39] Again, I guess that most of us would not consider this to be an essential effect on creativity, though.

The third question, then, is, if the cannabis high can have a *characteristic* creativity-enhancing effect. Let's assume that cannabis can have a substantial as well as an essential effect on our creativity by allowing us to better freely associate and connect far-remote concepts and ideas. Could the same effect be achieved with micro-dosing LSD? What is the characteristic effect of cannabis on creative activities as opposed to the influence of other mind-altering substances or techniques like meditation?

Again, let's modify our initial question about cannabis and creativity:

"Can a cannabis high temporarily enhance creative activities in certain phases substantially, essentially, and in a characteristic way?"

I hope that at this point it has become visible how much it matters to ask the right question(s) if we are looking for satisfying answers.

On a practical level, many cannabis users would be satisfied to know whether cannabis can lead, for instance, to relaxation or a focus of attention to substantially help them in a creative phase. They probably do not even ask for an essential effect in the sense defined above. If cannabis can bring a substantial contribution to their creative output, especially in a situation where it is hard for them to find relaxation by going on a three-week vacation to the Seychelles, they may be totally satisfied.

Others, however, may find it more interesting to know whether cannabis can bring something more essential to their creative abilities – and maybe, also, if cannabis has a characteristic effect spectrum on creativity, an effect pattern that may or may not work better for some creative activities than other mind-enhancing substances or techniques.

In the following, I will argue that cannabis users have reasons to believe that a cannabis high can have substantial as well as an essential positive effect on creative activities – especially if used with knowledge and skill. Also, I will outline an answer to the question in which way a cannabis high may have characteristic effects on our creativity. I will also explain why I think that cannabis can also be detrimental to creative activities given various conditions. Before, however, let me comment on the nature of a few recent scientific publications on cannabis and creativity which generated much interest in the public.

Cannabis and Creativity: Scientific Investigations

There are only very few recent scientific studies concerning cannabis and creativity. [40] [41] Studies interpreting them usually concentrate on the concepts of *divergent thinking* and *convergent thinking* for creativity because there are standardized tests for these cognitive abilities. *Divergent thinking* can be assessed for instance with the *Alternate Uses Task (AUT),* where individuals are asked to generate, as many as possible, uses for a certain object, for instance, a book. Some answers would be that you can read a book, help ignite a fire with the pages, kill a fly with it, step on it to reach something high up, etc. Your answers are then evaluated along the lines of certain criteria, for instance, originality.

Convergent thinking is usually measured with the *Remote Associates Test (RAT)* in which people are asked to find a concept that can be meaningfully related to three seemingly unrelated concepts, e.g. "widow", "bite", and "monkey". The solution here would be "spider".

Luisa Prochaskova and Bernhard Hommel give a short description of the two styles of thinking tested with these two tests, respectively:

"In sum, whereas divergent thinking calls for an associative, parallel, and flexible type of processing, convergent thinking rather calls for analytical, serial, and persistent processing to converge on a single answer." [42]

There is some evidence, by the way, that natural synesthetes seem to perform better than others in verbal divergent thinking as well as in visual convergent thinking tasks.[43]

In his *Metacontrol State Model* (MSM), psychologist Bernard Hommel argues that all creative performance involves a *balance* between persistent information processing (convergent thinking) and flexible processing (divergent thinking) and that this balance is guaranteed by metacontrol states. I will not go into details here concerning Hommel's MSM, but I want to highlight two core implications of it with which I agree:

(a) Creativity is not just divergent thinking or one other uniform mental process

(b) For successful creative performance, we have to be able to consciously or subconsciously find a balance between various cognitive processes

The few studies on cannabis and creativity we have seen so far may be an interesting start to understanding better how cannabis could affect some cognitive abilities that play a role in creative activities. We need to keep in mind, though, that a cannabis high comes with a complex bouquet of effects on perception and cognitive and perceptual processes such as attention, memory retrieval, the perception of time, and other mental processes. Importantly, also, it needs to be stressed that various creative activities each involve the concertation of dozens of different cognitive, perceptual, and maybe even motor control skills – as we have seen in the two examples of the creative basketball player.

Most commentators on cannabis and creativity

referring to some small scientific studies out there tend to vastly overstate the implications of these studies on the complex broader issue of how cannabis may affect creativity.

Let me focus on two problems regarding studies that have been done so far: First, they underrate and misrepresent the wide bouquet of cognitive effects described by cannabis users. Second, these studies often seem to work with a simplistic understanding of creativity, equating creative performance with divergent thinking, for instance.

In his article "What's wrong with creativity testing", Robert Sternberg summarizes his criticism of such a simplistic view of creativity:

> "There is no reason to believe that the different kinds of creativity represent, simply, different amounts of a single unidimensional construct. (...) The point of view presented here does not suggest that current creativity tests are invalid, but rather, that care must be taken like claims made for them."[44]

Like Sternberg, I believe that we can learn something from these studies. But we need to do a lot more work to better understand how diverse creative activities can be and how cannabis can influence these activities.

Let me highlight another typical flaw in some of the scientific studies we have seen so far. These studies look at various standardized cognitive test performances of subjects *during* the acute influence of cannabis. As we have seen, however, the altered state of the cannabis high is characterized by a whole *bouquet of cognitive effects* which can lead to *cancellation* effects as described above. These cancellation effects, however, depend on the skills and knowledge of users and environmental and other factors. They can be avoided by users if they, for instance, learn about using cannabis creatively only for certain phases of the creative process or if they learn to dose cannabis individually for various phases or processes.

One example would be a strategy of cannabis enhancement for creative purposes which I named a "deep dive" during a strong high: The cannabis user lays down during a strong high to generate, for instance, some intense mental imagery during an enhanced state of imagination.[45] She is probably not able to get into a sophisticated creative activity like writing a short story well during such a strong, associative high. So, she just waits until her high is weaker to use the remaining enhanced focus of attention to get into the process of writing. In this second, more analytical phase, she can then use her memory of the creative vision she had in a previous phase. A good strategy would be to take short notes during the high. In this way, cancellation effects could be avoided, or at least minimized. Studies that focus on cognitive effects during a high with their standard testing methods cannot take such a strategic creative enhancement into account.

One famous creative artist who successfully employed something like a "deep dive" technique was the Spanish painter Salvador Dalí, who used his dreams and self-induced states of lucid dreaming as techniques to enhance his creative work as a painter. He once said:

"Give me two hours a day of activity, and I'll take the other twenty-two in dreams."

Dalí would probably not have scored well in a RAT or AUT test while he was in a state of lucid dreaming, but many of us value the creative outcome of his efforts.

When people go to ayahuasca retreats, they usually

have somebody there to guide them through integration sessions, sessions that come after the actual "trip". In these sessions, they learn how to integrate what they have experienced into their lives and make some sense of it. Similarly, we can use the cannabis high and learn techniques on how to maybe keep short notes and later use our experiences during a high for creative (or other) purposes.

(9) The leaf of a Silver Haze plant, allegedly a hybrid with a strong sativa-dominant heritage. Silver Haze is a further breeding evolution of the breeding classic Haze, a mixture of equatorial genotypes considered by some connoisseurs as the purest and strongest Sativa. The leaves of Sativas are said to be more slender and serrated than those of Indicas. There have been various scientific debates about the Sativa/Indica and Ruderalis distinction itself and whether these are species or subspecies of cannabis.

Suggestions for Future Scientific Investigations

There are more ways for scientists to proceed when it comes to investigating questions concerning cannabis and creativity. The most promising would be to take a closer interdisciplinary look at the complete bouquet of perceptual and cognitive effects of the cannabis high, to find those neuronal correlates, and to explore how these effects are interrelated.

Certainly, this needs to be an interdisciplinary effort informed by more basic research into the endocannabinoid system (ECS) as it is involved in higher cognition. I will not go into the whole list of disciplines of cognitive effects here again. Let me mention only a few cognitive effects and enhancements during a high that cannabis users have described over and over again: a better mental focus during a high, an enhanced ability to perceive 3-dimensions in 2-dimensional imagery, a strong feeling of awe and curiosity, an intensification of sensory experience, a redirection of their attention, for instance, to their body, an enhanced ability to perceive bodily states, synesthetic experiences, a better ability to retrieve distant memories in detail, enhanced ability for pattern recognition, a more intense imagination, and enhanced introspection, mind-racing, and an enhanced ability to better empathically understand others.

Note, importantly: these effects could each in themselves enhance a creative activity if used by a knowledgeable cannabis consumer. All of them could potentially have substantial effects on creativity, and we would think of some of them as having influences on cognitive abilities that are essential to creative processes. So, even if science can confirm only one or a few of the cognitive effects described for users of a high, such as the enhancement of our ability to imagine situations, this would suffice to claim that cannabis has the potential to be creativity-enhancing. Many cannabis users have experienced enhancements such as the intensification of imagination so often that they will not want to wait for scientists to jump in to tell them they are right – in some decades.

On the other side, many users have also reported potentially detrimental cognitive effects such as losing the thread during a conversation, the distortion of their sense of time, or paranoia due to the overinterpretation of patterns during a high. Only if we understand better how the endocannabinoid system is involved in these processes and how cannabis and cannabinoids can affect these perceptual and cognitive processes as a whole will we learn more about the potential of cannabis as a creativity-enhancing tool, as well as how it can negatively interfere with various creative activities.

For scientists, this is a long and rocky path with a lot of conceptual and empirical groundwork to do. Understandably, it is tempting to take a shortcut and to rather create a study based on observations about cannabis users and how they behave during a RAT or AUT test. And it is tempting for journalists and other commentators to pick up the headlines of one of those studies to come up with a verdict on cannabis and creativity: *"Scientists show that cannabis negatively/positively influences creativity!"* Lovely clickbait for sure, especially in a situation where you want to seem like you are doing scientific reporting. But the grandiose claims in those headlines, as so often in science reporting, are not warranted. Divergent thinking alone is not creativity and preliminary studies with the specific method of administering tests during a high do not show us much about the full potential of the cannabis high to enhance our creative activities.

Characteristic Effects of the Cannabis High

Can we say more to cannabis users about whether a cannabis high characteristically affects their creative abilities in various situations? Do we have reasons to believe that cannabis could have a characteristic effect on various creative activities?

Let me outline the beginning of an answer to this question. Cannabis brings a "signature" bouquet of cognitive effects which characterize a multidimensional cannabis high state. This signature, as far as I can see, is very different for various cannabis varieties and depends also on factors like the method of consumption, mood, and setting, but, arguably, there are similarities between those "highs". Note, however, that some factors may well systematically shape the "signature" of a cannabis high as it comes to mind enhancements. Aged cannabis, for instance, contains more Cannabinol (CBN), a degradation product of THC), and other degradation products of cannabinoids and terpenes which are believed to cause sedative effects. These may also be partially responsible for problems with your short-term memory during a high. If this is so, then we would have to expect a lot of "cancellation" effects from aged cannabis. If you are tired and you tend to forget about the great idea for a mathematical theory you just had a few seconds ago, then you will probably never end up winning the Fields Medal for the theory that got lost in time.

Individual Reactions and Subjective Flow

The good news for cannabis users is that we now know

that cannabis is a remarkably nontoxic substance and that it may be well worth a try to use for creative purposes. Furthermore, there are some easy methods and tricks based on what we know about how we could avoid some pitfalls and how we could improve our use of cannabis for creative purposes. Generally, a mindful approach to experimenting with cannabis can help you to find a dose that works for you. If you know more about creative phases and activities you can learn how to use specific doses of cannabis for various phases in various activities. For instance, after a mindful journey trying various cannabis varieties, you might end up using a cannabis variety in a high dose for a strong high in a phase of the music composing process in which you want to enhance your ability to imagine sounds and to come up with new ideas for sound combinations. Later, in the editing phase, you may choose to only use a light buzz to be able to vividly remember what you have experienced before to then review the ideas and make changes. Importantly, this mindful process of finding a method of consumption, the right variety of cannabis, a productive set, and setting is *highly individual*. We all have different talents, characters, and styles of perceiving the world, and we, therefore, have to find how a cannabis high can help us in various situations, and how it can be detrimental.

It is of crucial importance, therefore, that you trust your judgment more than the marketing of cannabis-producing companies when it comes to the effects of various varieties on your creativity. We have gone from the era of prohibition to the era of cannabis marketing, and a lot of the marketing claims about "Sativas" and "Indicas" are not warranted and are actually refuted by several scientific investigations.[46]

From a subjective point of view, it may be worth asking yourself what helps you to *"get into the flow"* of creative activity. A high may sometimes help you to get into the flow and to stay in it, but it can also let you drop out or even hinder you to get into a flow at all. The Hungarian psychologist Mihály Csíkszentmihályi described flow as

> *"(...) being completely involved in an activity for its own sake. The ego falls away. Time flies. Every action, movement, and thought follows inevitably from the previous one, like playing jazz. Your whole being is involved, and you're using your skills to the utmost."[47]*

According to Csíkszentmihályi, if we want to achieve flow experiences, we have to achieve some kind of *balance*: If the challenge of a creative task is too ambitious concerning our skills, we might become frustrated, angry, or stressed and we do not achieve our goals. If the challenge is not enough for our skills, we will feel bored, distracted, and maybe even depressed at some point.

This takes us back to the surfboard metaphor. Like a surfboard, cannabis is a tool with a unique shape that can essentially and substantially enhance our lives. But we have to learn how to use it and how to control a whole range of factors that influence our performance. The board can be a great tool for a certain activity, but it cannot guarantee a positive outcome by itself.

The Persistent Mindful Surfer (PERMIS) – Approach

The surfboard is a very useful metaphor especially as it concerns a point that relates to Hommel's metacontrol states model (MSM) of creativity, for which *the control of a balance* plays a crucial role in the creative process. For a creative to use a high – or any other altered state of consciousness – productively and successfully, she will have to learn how to control various influential factors in her experience before she enters the high, during the high, and, then, after the high, to integrate her experience and to be able to remember it and use it productively. Let's call this the Persistent Mindful Surfer (PERMIS) approach to using an altered state of consciousness for creative purposes. It involves:

Attitude of Persistence

An attitude of goal-directed persistence towards using your altered state specifically for creative purposes.

Creative Skills and Knowledge

If you can't play piano, then the impact of an altered state of consciousness will be limited as to improving your ability to produce a great piano solo, of course. The enhancement of your creative output with an altered state of consciousness, then, naturally depends on your previous skills needed for a creative activity, including general creative techniques such as brainstorming, etc. [48]

Knowledge about Integrating the High

Previously acquired knowledge that helps you to use the altered state for creative activities, especially knowledge about positively influencing factors such as dosing, will affect the spectrum of the psychoactive substance used. The better your information about an altered state, the more you will be able to use it as a tool for creativity.

Skills and Knowledge Riding a High

Importantly, you will have to learn how to best maintain a creative flow during the experience to ride your altered state of consciousness like a wave. This involves a deep understanding, if only intuitive, of how to navigate and keep a certain balance during an altered state of mind. This involves a high level of introspective control, which means that you have to be able to constantly perceive your mind in the flow during the experience. These skills can be trained with mindful experimentation. Importantly, this knowledge is highly dependent on mindful self-knowledge: am I ready for this strong dose in this specific situation? Is it the right mood and environment for me to use this variety of cannabis? Can I still attend to my environment enough to be able to perform creatively during the high as it unfolds in my mind?

In my book, *The Art of the High. Your Guide to Using Cannabis for an Outstanding Life*, I give some more elaborate advice based on our state-of-the-art knowledge to use cannabis for creative and other mind-enhancing purposes.[49]

Summary

We have seen that the question *"Does cannabis enhance creativity?"* is misleading, vague, and too simple to interestingly guide an investigation on the impact of cannabis on our creativity. It ignores the fact that cannabis is like a tool: a tool cannot guarantee a positive outcome, it can only be useful in certain environments for certain tasks if used with skill and knowledge.

After some reflection, we ended up with a more interesting question:

"Can a cannabis high temporarily enhance creative activities in certain phases substantially, essentially, and characteristically?"

As we have seen, this redesigned question shows that for future scientific investigations, we have to design studies differently – and we certainly have to do some more work to answer more questions:

What do we consider to be the *essential* cognitive abilities that define various creative activities?

Are these cognitive abilities different for various creative processes?

How could we avoid "cancellation" effects in our study designs? Etc.

Scientists should not be satisfied with the studies specifically on the subject of cannabis and creativity that have been done so far. I have argued that there is a growing body of evidence from various sources beginning to show us a signature pattern in which the cannabis high can typically alter a whole bouquet of cognitive functions. Cannabis users can learn how to mindfully experiment with cannabis to avoid the cancellation effects of cannabis for creative activities and to more productively use a high for creativity. Given the experiences of users and other empirical knowledge so far, we have good reasons to believe a cannabis high indeed has a great potential to enhance various kinds of creative activities.

(10) The leaf of a Snow Ryder, also an auto-flowering variety. It is supposed to be strongly Indica-dominant variety, which is reflected in the rounder shape of the leaves. In the last years, however, studies involving genetic profiling of existing plant varieties on the market today showed no significant characteristic difference between varieties labelled as "Sativa" or as "Indica". Also, the hybridization of cannabis varieties seems to be at a point now where the plant morphology usually does not consistently indicate anymore a certain genetic heritage.

Epilog I: Cannabis, Creativity, and Culture

Let me shortly address a different, but related question: How much have cannabis users influenced our modern human society with their cannabis-inspired ideas and work?

There are many highly influential philosophers, writers, scientists, artists, architects, musicians, business people, comedians, and others who have productively used cannabis, some of them explicitly for their work:[50] writers Robert Louis Stevenson (*Treasure Island*), Charles Baudelaire (*Les Fleurs du Mal*), Alexandre Dumas (*The Three Musketeers*) , Victor Hugo (*Les Misérables)*, Rudyard Kipling (*The Jungle Book*) Marcel Proust (*A La Recherche Du Temps Perdu*), William Butler Yeats (*Collected Poems*), and Mark Twain (*Adventures of Huckleberry Finn)*, Jack London (*The Sea-Wolf*), philosophers Walter Benjamin (*The Work of Art in the Age of Mechanical Reproduction*) and Ernst Bloch (*The Principle of Hope*), writers Jack Kerouac (*On the Road*), Alan Ginsberg, the architect Frank Gehry, Norman Mailer, psychologist Timothy Leary, writer Robert Anton Wilson (*Prometheus Rising*).

The astronomer and popularizer of science Carl Sagan (*The Dragons of Eden*), used cannabis and used it explicitly for its mind enhancements; filmmakers Robert Altman (*M*A*S*H*), Francis Ford Coppola (*Apocalypse Now*), and Hal Ashby (*Harold and Maude*), Nobel Prize-winning physicist Richard Feynman, Barbara Ehrenreich (*Nickel and Dimed: On (Not) Getting By in America*), Nobel Prize-winning molecular biologist and discoverer of the structure of human DNA Francis Crick, neurologist and bestselling author Oliver Sacks (*The Man Who Mistook his Wife for a Hat*).

As to comedians: Groucho Marx was named after a "Grouch bag" in which kids in the 1920s would keep their pennies, pieces of candy, and their marijuana.[51] Groucho Marx from the Marx brothers, who once in a movie pointed to a corpse saying, "Either this man is dead, or my watch has stopped". Other comedians who used cannabis for inspiration were Lenny Bruce, George Carlin, Steve Martin, and Bill Hicks ("Why is marijuana against the law? It grows naturally upon our planet. Doesn't the idea of making nature against the law seem to you a bit paranoid?"). Bill Murray and Kat Williams are two more on a list that could be extended almost endlessly.

We could fill a book only with prominent names of cannabis-using musicians: Cab Calloway, Bessie Smith, Louis Armstrong, Mezz Mezzrow, Billie Holiday, Lester Young, Gene Krupa, Bob Dylan, Art Garfunkel, Bob Marley, Peter Tosh, The Beatles, Jerry Garcia, Jimmy Hendrix, Janis Joplin, Patti Smith, Gilberto Gil, Willie Nelson, Ben Harper, Lil Wayne, are certainly only a few examples of a much longer list of musicians who have used cannabis. The same can be said for actors, including Errol Flynn, Robert Mitchum, Tony Curtis, Johnny Depp, Jack Nicholson, Susan Sarandon, Natalie Portman, Kirsten Dunst, and Frances McDormand, to name just a very few.

No doubt their work had a huge influence on our culture. Yet, how much did cannabis play a role for them and their ideas? Could they have done the same work or even come to better results without the use of cannabis? That is an entirely different question. We would need to delve deeper into the biographies of those luminaries to come to a conclusion about the impact of cannabis on their ideas. I have begun to do this in my essay collection *What Hashish did to Walter Benjamin*, in which I explore how

Benjamin, the philosopher Ernst Bloch, the astronomer and popularizer Carl Sagan, jazz musicians like Louis Armstrong and Billie Holiday, and many others have successfully used cannabis to find inspiration for their work [52]

If we look at the larger picture of societies and the impact of cannabis use, it is yet another question whether so far, cannabis has played a positive role that concerns the total creative output of modern societies. In the last decades, the prohibition and ensuing lack of information and education on cannabis had a negative impact and led to more individual "cancellation" effects, thereby diminishing the creative potential of cannabis for society. So, even if cannabis inspired some to do brilliant creative work, it may have had a negative impact on many others.

In the light of what has been said before here, the following may be a more interesting question to ask about the cultural dimension of cannabis use for creativity:

What could cannabis (and any other psychoactive substance, or mind-altering techniques like meditation) do for the creative output of our society as a whole, if it would be accessible and mindfully used with care, skills, and knowledge?

Epilog II: Cannabis, Freedom, and Cognitive Liberty

In his famous article "On Liberty", philosopher John Stuart Mill claimed that a person's thoughts, feelings, beliefs, plans, and actions must lie within a protected domain of liberty:

"(...) the inward domain of consciousness; demanding liberty of thought and feeling, absolute freedom of

opinion and sentiment on all subjects, practical or speculative, scientific, moral, or theological . . . liberty of tastes and pursuits; of framing the plan of our life to suit our character; of doing as we like, subject to such consequences as may follow: without impediment from our fellow-creatures, so long as what we do does not harm them, even though they should think our conduct foolish, perverse, or wrong."[53]

Mill also argued in connection with alcohol prohibition laws that as long as a person does not harm another, he should have the right to inebriation.[54]

Mill's essay is often quoted when it comes to arguments against cannabis prohibition. I fully agree with his thinking and would like to add an important point to it. The freedom addressed by John Stuart Mill in this essay only covers one aspect of freedom, as he states himself in his introduction: *"The subject of this Essay is not the so-called Liberty of the Will, (...) but Civil, or Social Liberty: the nature and limits of the power which can be legitimately exercised by society over the individual.*"[55]

Now, let us look at another very important aspect of freedom. Our concept of freedom is intimately connected to our creative abilities. As a creative being, I do not only have the freedom to choose between a whole variety of given alternatives, like a post-modern consumer in a supermarket: my freedom importantly arises out of my ability to think and act creatively, to help create and transform the reality I want to live in and to become the person I imagine to be through a process of self-directed personal growth.

Creativity is the pulsating heart of freedom. Only an inventive mind can go beyond the choice of existing paths and freely create a new self and a new world.

If cannabis or any other psychoactive substance has the potential to enhance one of our most fundamental human cognitive abilities, to think and act creatively, then we should most fervently defend our right to use this substance to enhance our ability to truly be free.

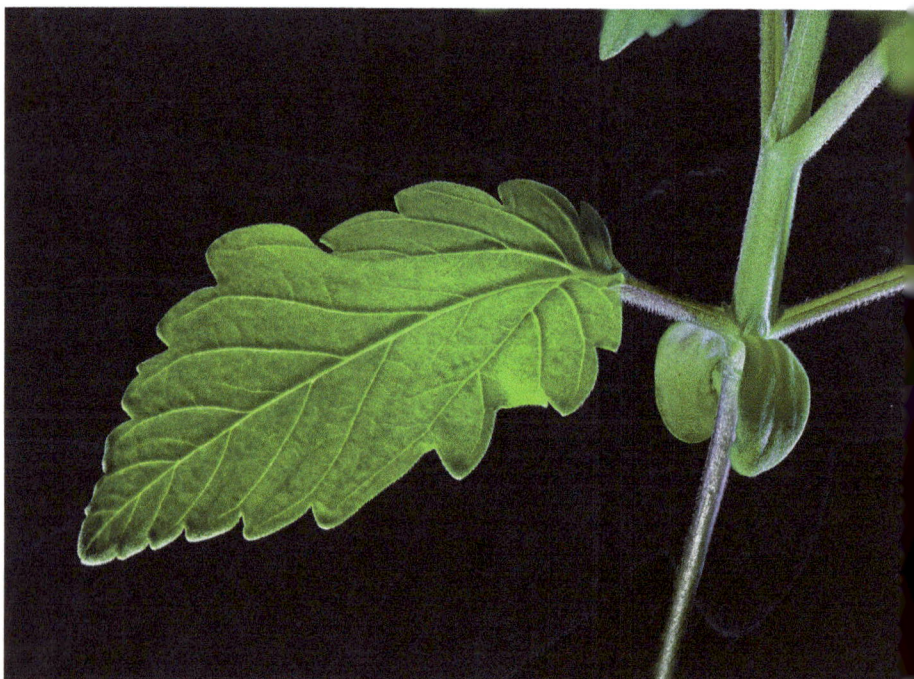

(11) The early leaf of a White Haze plant, also a further development of the classic Sativa variety "Haze". Cannabis marketing in 2022 still relies on labelling cannabis on Sativa and Indica, based on the paradigm that Sativas bring a more energetic, heady high, while Indicas affect your body more and tend to be more sedative. Clearly, however, even if these generalizations about Indica and Sativa genetics were useful and to some extent true some decades ago, it makes not much sense anymore to classify cannabis hybrids plants today with ratios of Sativa or Indica heritage, as mentioned before.[172]

IV. Cannabis Insights:
Myth or Reality?

Countless users of cannabis have claimed that a high can act as a catalyst to obtain profound and useful insights. Are these reports just exaggerations of users justifying their smoking habit? Or is it true that a cannabis high can lead to profound insights?

In his legendary essay "Mr. X" published in Lester Grinspoon's book *Marijuana Reconsidered* (1971), an anonymous author stated:

> *"There is a myth about such highs: the user has an illusion of great insight, but it does not survive scrutiny in the morning. I am convinced that this is an error, and that the devastating insights achieved when high are real insights; the main problem is putting these insights in a form acceptable to the quite different self that we are when we're down the next day."*[56]

With the permission of the author, Lester Grinspoon would reveal the identity only posthumously. It turned out that the article had been written by Grinspoon's best friend, the late Carl Sagan, famous astrophysicist and popularizer of science who died in 1996. In his essay, Sagan writes:

> *"I can remember on one occasion, taking a shower with my wife while high, in which I had an idea on the origins and invalidity of racism in terms of Gaussian distribution curves. (...) One idea led to another, and at the end of about an hour of extremely hard work,*

I found I had written eleven short essays on a wide range of social, political, philosophical, and human biological topics. I can't go into the details of those essays, but from all external signs, they seem to contain valid insights. I have used them in university commencement addresses, public lectures, and my books. "[57]

Carl Sagan is not the only prominent cannabis aficionado who used marijuana for inspiration. Pulitzer Prize winner Norman Mailer once said in an interview for *High Times* Magazine:

"(...) what I find is that pot puts things together. Pot is marvelous for getting new connections in the brain. It's divine for that. You think associatively on pot, so you can have real extraordinary thoughts. But the more education you have, the more you have to put together at that point, the more wonderful connections there are to see in the universe. "[58]

Many other prominent luminaries have used cannabis for inspirational purposes. But how much did cannabis help them to come to insights? As we have discussed before, maybe cannabis simply helped them to relax and to get happy to get their already talented creative minds started. Even if some of them stated that cannabis helped them to come to insights, they could be self-deluded about the inspiring powers of cannabis because of what we could call *forgetful glorification* – ideas during a high often seem to be profound and glorious revelations, just like taste experiences during a high become so much more intense.

The Nature of Insights

First, let us look at short definitions of the word

"insight" in English. There are rather different sets of cognitive processes and abilities denoted by this term if we take a closer look at how it is really used. Merriam-Webster defines insight as "the act or result of apprehending the inner nature of things or of seeing intuitively." I believe that this definition well summarizes what I would subsume, minimally, under the notion of insight as meaningful pattern recognition. Here is an example of this kind of insight from my own experience: I gained insight into the nature of the cannabis high by using a macro lens to take photos of the plant, seeing how trichomes and stigmas change their color and form during the process of maturation, which made me understand better how important it is to cure the plant and to consume it in the right phase of the curing process.

I discovered a visual pattern because I took a closer look and used some sophisticated equipment to get better perceptual access to something I had not been able to see this way before. Note that this kind of insight may be quite important and substantial to a person and lead to personal development. There is nothing much creative about this first kind of insight. I do not even have to perceive a new pattern. Maybe I have just used my normal pattern recognition, but this time, I had better perceptual access to the object of my curiosity.

Here is another example of this kind of insight: if you for instance have better access to episodic memories about your childhood and remember previously forgotten or suppressed traumas, or formative beautiful experiences, this act of memorizing could lead to rather substantial insights into your personality or in the personality of others. A cannabis high has often been reported to lead to an enhancement of episodic memory, so this could be one reason why many people report significant insights. It is also rather obvious that a simple redirection of our attention could lead to those kinds of insights, as we start attending to patterns we have not even taken into our focus before. Cannabis users have described how their attention was redirected for instance from listening to somebody talking about a certain subject to his body language. Suddenly, patterns of behavior appear that have gone unnoticed before, such as the attitude of pride expressed in a posture or hand gesture.

Other mental enhancements of the cannabis high could play an important role to bring about this kind of insight. For instance, the feeling of awe and the hyperfocus of attention, which may cause us to look at something longer, to perceive more details or new aspects in whatever we are perceiving. As we have seen in our discussion of *synesthesia*, there are good reasons to believe that a cannabis high may significantly help our ability to recognize new patterns. This ability may become especially valuable for us when we reflect upon our past behavior and emotions, moods, our tics, our development. Our enhanced pattern recognition could for instance lead us to understand that we are in the grip of certain habits, such as being defensive when criticized instead of really contemplating and listening well to reasonable critique. This insight could help us tremendously to de-condition ourselves from bad habits, a process that can be profoundly liberating. We will talk about this in more detail when discussing the topic of cannabis and addiction.

Again, then, this kind of gaining insight does not need to be what we prototypically see as creative – it is just a relatively sudden deeper understanding of the nature of something we are perceiving or thinking about. Note, by the way, that this kind of insight, like sudden creative insight, could well be accompanied by something like a *Eureka* "Aha"-experience, for instance when we suddenly realize that a childhood trauma has negatively shaped our behavior for years.

Usually, when cognitive scientists or psychologists investigate the nature of insights, though, they mean to clarify the nature of sudden *creative* insights or ideas. Interestingly, as we will see, they are mostly concerned with insights that we have when we want to solve problems that demand a creative, non-linear solution. The Cambridge Dictionary for instance defines insight as *"(the ability to have) a clear, deep, and sometimes sudden understanding of a complicated problem or situation."*

It is important, though, to distinguish these creative insights facing a certain kind of problem from spontaneous ideas or insights. I assume that we would all agree that in our everyday language, we use the term "idea" and "insights" to be something that not only occurs when we are thinking about a problem. I may have great spontaneous ideas or creative insights about a new technique of building a house even if I am not an architect or thinking about building a house for myself at all. Maybe I was just introduced to new materials and I come up with this idea. I have not even tried to solve a problem. My idea does not answer a certain problem or task, neither consciously, not unconsciously.

So, let me summarize at this point my observation about our natural linguistic use of the term "insight." I distinguished three different uses of the term insight:

(1) Insight is a sudden *deeper understanding* of the nature of a certain subject

(2) Insight as spontaneous, sudden creative problem solutions facing certain tasks that demand "out of the box thinking"

(3) Insight as spontaneous, sudden creative ideas which do not arise out of a consciously or even unconsciously perceived problem or task

When cannabis users report insights during a high, we have to take into account that they could mean all three types of insight identified here. I have argued above that given some of the mind enhancements of cannabis like hyperfocus, a redirection of attention, or episodic memory, it becomes plausible that a high could help for insight as in (1), a *deeper understanding* of the nature of something.

In what follows, I will take a closer look at what is known about the nature of creative insights and how a cannabis high may enhance our ability to produce these creative insights, both during creative problem solving as well as producing spontaneous ideas.

Let me make one more comment about the differentiation between *creative processes or activities* as we discussed them in the last chapter, and *creative insights* at this point. I assume that when we talk about creative insights we emphasize the intellectual nature of these creative processes. Briefly, then, I assume that we can act creatively without having a conscious insight that we could necessarily verbalize. Assume a painter intuitively starts using a different drawing technique for his oil paintings. Let's say that the outcome is novel and that we would also see it as valuable. I would say there is a difference in whether this creative activity was insight-driven or not. Maybe the painter was thinking about depicting daylight patterns on the water better and, therefore, came up with this technique. I would call this an insight-driven creative step. But it may have been just an intuitive creative leap that was much less "intellectual". Art critics may probably later reconstruct his new style as an insight, but that does not necessarily mean that the artist was fully aware of it under the description they would use.

(12) Cannabis leaves from a different perspective. In this close-up we can already see some of the mushroom-shaped trichomes in which the plant builds its cannabinoids, terpenes, flavonoids. Cannabis plant can express more than 200 terpenes, but most of them occur only in trace amounts. Other terpenes like for instance myrcene, limonene, and beta-caryophyllene, often occur in higher amounts and contribute to the characteristic effect spectrum of a variety. Therefore, some cannabis varieties for instance tend to be more sedative, while others produce a clear, energizing high. Cannabis varieties can also widely differ in their effects on mood and cognitive functioning, such as the way in which they affect our working memory, or our attention.

Insights in Gestalt Psychology and the Cognitive Sciences

So, let's take a closer look at creative insight and see how science approached this phenomenon. We all have those "Eureka" experiences and little or more profound insights once in a while. Suddenly, you realize that you can take a shortcut to drive to the supermarket. In an instant, you understand that one day you have to become an artist, or you realize in a flash that your marriage is going nowhere. Insights occur spontaneously, like sudden quantum leaps in understanding; rather mysterious, but sometimes with a profound impact on our lives. For a long time, mainstream psychology in the last century has persistently ignored this phenomenon. Insights seemed to be a myth bound to the classic Greek concept of a privileged genius receiving great ideas through the inspiration of the gods.

It was only at the beginning of the last century that a group of scientists around the German psychologist Max Wertheimer started to look for explanations for what he called "productive thinking". Wertheimer's famous book *Productive Thinking* was also the result of his discussions with Albert Einstein on creativity in 1916 and an attempt to explain how the most famous "productive" and creative thinker of all time came to his revolutionary ideas in physics.[59] Since then, psychologists and cognitive scientists around the globe have tried to further develop the existing models of the so-called "*Gestalt* school" in psychology to better understand insights.[60]

Wertheimer and his students wanted to study insight empirically, but they had a problem: creative insights seem to be an elusive phenomenon. They are hard to predict, come spontaneously, and seem to occur mostly unconsciously. Those lucky ones shouting out "Eureka" when having a profound and meaningful insight usually cannot tell how they arrived at their ingenious ideas. Insight often seems to come out of nowhere. How can you empirically study such an elusive and unconscious phenomenon? Wertheimer's solution to this problem was a problem. Or, better, a class of problems. Wertheimer and his students came up with specially designed problems, problems that would demand a creative solution, and problems that needed out-of-the-box thinking. This way, they could have subjects sit in a lab and observe them trying to solve problems. And by changing some of the conditions and the situation of those problem solvers, they could get insights into the nature of insights.

Now, remember again my distinction between creative insight in problem solving and creative ideas. Here we can see why scientists looked only at the former, not the latter; insights during problem-solving are easier to study. Therefore, scientific investigations do not look at the three kinds of insight I have described above, but only at a more "narrow" understanding of the term, insight in problem-solving. This is an important observation, as I will argue below. Remember a similar observation of mine in the last chapter when it comes to scientific definitions of creativity, where creativity is "narrowly" construed as either divergent thinking or convergent thinking because we have standardized tests for these. I have argued that because of this narrow focus, scientists often miss that creative activities are more than that and that therefore, we have to criticize scientific studies as telling us about creativity in a broader sense, which comprises a wide spectrum of different cognitive activities.

Keeping this in mind, let us take a look now at what Wertheimer and his students had to say about the nature of insight in creative problem-solving. One of the basic ideas of the *Gestalt* psychologists was that in the "*Eureka moment*", a thinker unconsciously "restructures" his perception of a situation and finds a new pattern, or, as they would call it, a new *Gestalt*. Let me explain this shortly by looking at one of the most famous experiments with an insight problem. The German psychologist Karl Duncker, Wertheimer's most talented student, invented the now-classic "Candle Problem", where the subjects are given a matchbox, a candle, and some thumbtacks and are asked to attach the candle to the wall.

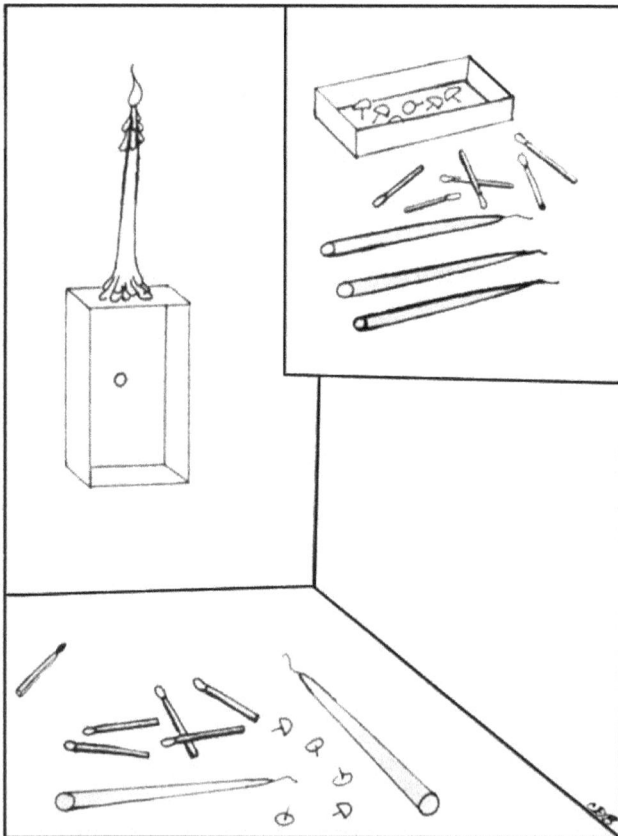

The candle is too thick to be directly tacked into the wall – the only solution is to use the tacks and attach the inside container of the matchbox to the wall and, then, put the candle on it. The crucial step for the problem solution is to see the matchbox container not as a container for matches, but as a tray for the candle. According to the *Gestalt* theorists, subjects trying to solve this problem need to *restructure* their first perception of the matchbox only as a container.

Duncker showed that the subjects needed longer to solve the problem if the matchbox container was presented to them with matches inside, highlighting the original function of the container. The subjects were blocked from coming to an insight because their perception and thinking of the box were "*functionally bound*"; they did not see the box as a mere object with a certain form that can be used in various ways but as a container with the specific function of holding candles. Duncker's candle experiment shows that a problem solver comes to an out of the box creative insight only if he overcomes this functional boundness by redirecting his attention to aspects of an object that he has not perceived before.

Let us assume that the subjects who solved the candle problem are faced with a similar problem a few months later. This time, they have presented needles instead of thumbtacks, a little cardboard box, and a little toy figure inside with a similar goal of attaching the toy to the wall. Subjects who succeeded in solving the candle problem would see the similarity of the solution and would easily solve the toy problem. Wertheimer would call this the use of "structural analogs": we transfer our knowledge from past similar experiences of problem solutions to new situations.

We now have three crucial concepts to characterize some important steps in the process of insight in problem-solving:

(1) *restructuring a problem representation* (where we have to perceive a situation differently), thus overcoming

(2) *functional boundness* (where we are bound to perceive or think of certain things as serving a certain function only), and

(3) *finding structural analogs* (or, in other words, finding similarities between patterns).

I will now use these three notions to give a rough explanation of how a cannabis high could affect our ability to generate insights in problem-solving.

Cannabis-Induced Insights

Users have described many effects of cannabis high on their consciousness, but for the sake of brevity, I will name only four here. One of the most common effects of marijuana during a high is on attention: stoners tend to have a stronger focus on attention, they *hyperfocus*. Sometimes they hyperfocus on sensations, a focus that leads to a more intense experience of the here-and-now. Sometimes they hyperfocus on memories or imaginations or a stream of thought. Second, they often redirect their attention, perceiving different aspects of a situation, such as redirecting one's attention from a TV show to a strongly intensified bodily feeling of stiffness in the neck. Third, many users have reported an *enhanced episodic memory,* i.e. an enhanced ability to vividly remember past events in their lives. In an article about the philosopher John Stuart Mill, N.S. Yawger wrote in 1938:

"John Stuart Mill . . . wrote of (cannabis') power to revive forgotten memories, and in my inquiries, smokers have frequently informed me that while under the influence, they are able to recall things long forgotten."[61]

Third, many users describe an enhanced ability to find structural analogs, or, similarities in patterns ("The jazz saxophone player Eric Dolphy sounds like the early Coltrane on this record"), ("This painting looks like an early Edgar Degas").

There are many other effects of a cannabis high that may positively affect our ability to gain insights, but for sake of brevity, let us, for now, stick to these four effects: *hyper-focusing*, the *re-direction of attention, enhanced episodic memory*, and an *enhanced ability to find structural analogs*. These will be enough to give us a rough outline of how marijuana can positively influence our ability to produce insights while high.

Here is what I take to be a very typical report of a marijuana insight:

"Martha found herself smoking (marijuana) with (...) Alice: During the conversation with Alice and Karl, I realized that she was being very self-conscious, and kept stepping back out of herself. I looked at her and thought, "That's a whole new way of looking at Alice." I had never seen her insecurities so palpable before. (...) I suddenly understood that her insecurity was a key to her personality, and then I also understood how it was a big key to my own, as well. I understood, too, how she and I clashed because both of us are insecure and that each of us was waiting for the other to give the cue of reassurance that never came. That's the type of insights I get when I am stoned, and for me, it's very useful."[62]

Under the influence of cannabis, Martha's attention is more focused on the here-and-now. She is intensely watching Alice's behavior. She has redirected her attention from understanding what Martha says to her body language. Her attentional focus has suddenly changed. Usually, she would probably be sharing her attention more, thinking about the content of the conversation, maybe listening to music, and keeping better track of time to make preparations for dinner for her guests. Her high makes her hyperfocus on certain patterns in Alice's behavior. Martha's attention is not *functionally bound* anymore to merely following the literal content of the discussion or catering to her guests, as it would normally be given the social etiquette of the situation. She is now open to redirecting her focused attention to perceiving a certain pattern in Alice's behavior, namely, behavior that signals certain insecurity. In the words of *Gestalt* psychology, she has "restructured" her perception of Alice. She can see a different pattern *Gestalt*, or, pattern in Alice's behavior and character now.

The change in attention during her high is not the only factor that allows Martha to proceed to her insight. For Martha to see this pattern in Alice's behavior, she must be able to recognize overall behavioral clues as fitting a pattern she already knows. Martha's *enhanced episodical memory* during her high might help her to remember past events in which she has seen Alice acting insecure; also, her enhanced episodic memory may help her to compare Alice's behavioral clues, such as avoiding eye contact, to similar behaviors she has seen throughout her life and learned to perceive as signs of "insecurity". Furthermore, Martha has to rely on her episodic memory to infer that she is insecure herself; she needs to remember episodes of herself showing that pattern. The insecurity behaviors that

Alice has seen or acted out herself before may be similar to that of Alice now. Last but not least, Martha's enhanced ability to see similarities between patterns allows her to see a "structural analog" between other behaviors of insecurity and some behavior she sees now in Alice. This allows her to understand that she and Alice are both insecure and therefore clash as personalities.

Of course, this is only a possible explanation of how Martha's insight occurred. Martha's report is not detailed enough to pin down how exactly she arrived at her insight. But the above possible explanation helps us to understand how the interplay of various effects of cannabis may positively influence Martha's ability to gain an insight into the background of the *Gestalt* model of insights.

Is Martha *solving a problem* creatively? Maybe, yes. Maybe she was responding to the conscious or unconscious tensions between her and Alice, which lingered in the back of her mind as a problem. This problem could have guided her interest in finding new insight as to why they clash. This is speculation, however. Her insight about Alice and herself could have also occurred without being consciously or unconsciously occupied with a problem. This brings us back to the distinction of insights as to the result of confronting certain problems or as spontaneous ideas.

When it comes to the enhancement of creative activities, I have argued that some of the effects of the cannabis high may cancel out some of the potential enhancements it brings. If I have a great idea during a high but fall asleep and forget it, or if I have a great idea for a trumpet solo while playing live, but cannot keep my timing as I should, some effects of a high could cancel out a potentially beneficial effect on creativity. These cancellation effects, arguably, play a big role when I am trying to actively solve a problem. Solving a certain complex problem brings some constraints that you do not have to face when producing a sudden idea. Let us say you are very high and your thinking becomes highly associative and "jumpy". You started thinking about a certain problem in evolutionary theory, but then, because of the appetite-inducing effects of your cannabis variety, you jumped to thinking about how hungry you are, you vividly imagine putting the remaining mango sauce on your peanut butter chocolate bar, you try and taste it, later develop a lovely idea for a dessert which will make you rich. If a scientist looks at how you solved the evolutionary theory problem, you do not fare well. But you generated a lovely idea.

Let me sum up then. Can a cannabis high lead to significant and deep insights? I have tried to show that it can do so, for all the three types of insights distinguished here: Insight as in a sudden deeper understanding as well as for insight in creatively solving problems as well as for ideas that do not come out of the mental occupation with a perceived problem. Again, we have to expect that there are cognitive cancellation effects, and these are probably more severe when it comes to insight in problem-solving with its strict constraints on what we would consider a successful insight. I suspect that therefore if a scientist would want to study the phenomenon of insights during a cannabis high with their classic experimentation style of confronting test persons with insights problems, they would come to rather negative conclusions.

The Long Path to Evidence

As I have argued before, the path to better evidence would be to investigate if a cannabis high can indeed lead to the various cognitive effects which could potentially play beneficial roles in the process of insight, such as hyper-focusing or enhanced episodic memory retrieval.

Modern neurocognitive theories explain insights as coming out of an interplay between different cognitive styles of the left and the right brain hemispheres. I have argued elsewhere[63] that it might be indeed interesting to see if cannabinoids may have a systematic effect on this interplay, as many enhanced functions during a cannabis high seem to be more right-hemisphere based. Intelligently executed neuroimaging studies of cannabis users, then, could help us to come to a deeper understanding of how a cannabis high may affect our ability to produce insights.

But we need to be realistic. First, even though we are seeing the legal situation for medical use as well as for adult use of cannabis changing around the world, the long decades of prohibition left us with many myths and disinformation about cannabis and its alleged toxicity. Science is driven by pragmatic interest, and many scientific investigations of the past have been mainly guided by the interest to find out about the risks of cannabis consumption rather than its potential. When it comes to research concerning cannabis as a mind enhancer, there is not all that much to draw on. Research needed to find out more about how a cannabis high could stimulate insights would be basic research – expensive, difficult to design, and hard to monetarize.

What I have tried to achieve here is to arrive at a deeper understanding of why so many cannabis users have reported in detail that they had profound insights during a high. I have analyzed the use of the common-sense term "insight" with the tools of an analytic philosopher and tried to come to a better understanding of what our everyday notion of "insight" actually denotes. I have then looked into some of the research from cognitive sciences and psychology to come to a better understanding of how various cognitive effects of a cannabis high could affect our ability to obtain insights.

I hope I have broken down the question about insights during a high to something that can be studied more easily. We should take the many detailed anecdotal reports about various mind-enhancements during a cannabis high seriously and start to investigate them empirically. Once we better understand the effects of cannabis on our episodic memory, our attention, and other cognitive processes, we will be able to get a better picture of how these changes in cognition can lead to more complex enhancements of insights under favorable conditions.

So far, the overall anecdotal evidence seems to me to be highly interesting and to some degree supports the hypothesis that a high can be a useful tool to obtain insights. We have many independent detailed reports about stoner insights, and many more independent and detailed accounts of a whole spectrum of other cognitive effects during a cannabis high – effects, which, as we have seen, are potentially crucial for the generation of insights. It becomes more and more clear now that our ability to produce creative insights crucially depends on right hemisphere activity, a fact that Carl Sagan had already addressed more than 50 years ago in a footnote in his *Dragons of Eden*, where he speculated that a cannabis high might suppress left hemisphere activity in favor of right hemisphere processing. Surely, further research in this area will be both fruitful for the understanding of cannabis effects as well as for our understanding of creativity and insights in general.

For a long time, our perception of cannabis has been *functionally bound* too much by focusing on its risk potential. Do not get me wrong: I do not want to argue that we should ignore any risks or negative aspects of cannabis consumption. Clearly, there are risks, and obviously, the value of insights could be diminished or even completely canceled if we do not take a mindful approach to cannabis use. If I have a lovely insight during a high and do not note it down to later remember it, it may just get lost in time, like a tear in the rain. And if I have deep insights about the nature of black holes during a strong high while driving, it may diminish my alertness and kill me and some others. Insights are generally an elusive phenomenon, and insights during a high need even more of an intelligent and mindful approach to become useful; those butterflies are hard to catch. Also, because of what cognitive scientists have called state-specific memory: in a phase of depression, we tend to better remember sad times, while in a state of joy, we better remember happy times. If we are high, we better remember times when we are high. Insights during a high often need to be elaborated and worked out in a more sober state to be valuable to us, so we need to be mindful, and we need to change our attitude towards a high, assuming that it can bring highly valuable ideas and insights.

Scientists in the field of consciousness research and those interested in the psychological effects of cannabis need to free themselves from the still predominant one-sided fixation on the risks to get a deeper insight into the positive potential of cannabis as a mind-enhancing tool; a potential that so profoundly and positively affected millions, if not hundreds of millions of users – and through some of them and their work, the history of cultures worldwide.

(13) A male cannabis plant in the flowering stage. Male plants contains much less of the main psychoactive Delta-9-Tetrahydrocannabinol (THC).

V. Cannabis, Introspection, and Personal Growth

"Once again the powers of the herb open up the mind. Seek deep inside. Tell me what you find."
Cypress Hill, Temples of Boom III

As we mature, we have experiences and gain knowledge about a wide variety of situations and facts. Accumulating knowledge about the world lets us grow to become students, teachers, professors, mothers or fathers, skilled professionals, or masters of some kind. However, there is a special form of knowledge that shapes us the most as we grow: self-knowledge. Our self-knowledge plays a central role in who we are, how we lead our lives, how we treat others, how we deal with problems, and how we personally grow. You may remain more or less the same person as you learn new facts about the moon landing, the behavior of red ants, or Fibonacci numbers, but once you learn that your inability to lead a happy marriage is caused by a trauma in your childhood, this one piece of knowledge could change the course of your life forever.

The Ancient Greek imperative "Know yourself" inscribed in the forecourt of the Temple of Apollo at Delphi unquestionably confronts us with a task that will never be completed. Nonetheless, the fact remains that we all know how rewarding and life-changing this path can be. Myriad consumers of cannabis as well as consumers of other psychoactive substances like LSD or psilocybin have reported instances of important insights into themselves – insights that significantly led to augmenting their self-knowledge, and, consequently, to their personal growth.

Based on his questionnaires sent to 750 consumers, Harvard psychologist Charles Tart found in his 1969 study *On being Stoned* that many of those who responded agreed with the following description of the effects of a cannabis high:

> *"Spontaneously, insights about myself, my personality, the games I play come to mind when stoned seem very meaningful."*[64]

But how could cannabis be especially helpful in gaining self-knowledge?

"Introspection" in Philosophy and Common-Sense

We usually consider introspection to be the main route to self-knowledge. The term introspection is a composite of the Latin words *"spicere"* ("to look") and *"intra"* (within). The metaphor suggests that we see or (visually) perceive our inner mental processes. Obviously, however, this should not be taken too literally: neuroscientists did not find eyes, ears, or other known sensory organs in our brains, and we certainly don't expect them to do so in the future.

So, what is introspection, generally speaking? Let us look at some perspectives from the modern philosophy of mind on this issue. *The Stanford Dictionary of Philosophy* states:

> *"Introspection, as the term is used in contemporary philosophy of mind, is a means of learning about one's own currently ongoing, or perhaps very recently past, mental states or processes."*

Our common-sense notion of "introspection", however, seems to be much broader construed: we usually say that we can introspect not only mental states like a current pain sensation, feelings of anxiousness, or a joyful mood, but we often say that we introspect dispositions (like for instance a tendency to overreact to critique) or other aspects of our personality like character traits. This broader view of introspection is expressed in the following statement by Pete Brady, a contributor to Lester Grinspoon's website *marijuana-uses.com*, who mentions an enhancement of introspection under the influence of marijuana:

> *"The marijuana high made me introspective, and I used it to catalog my strengths, weaknesses, and traits. The drug was a revealer, not an escape mechanism; it helped me see who I was and what I needed to be."*[65]

Another anonymous contributor ("Twinkly") to Grinspoon's website reports:

> *"I was so much more tuned-in to myself and others. I could concentrate on my fears, my turmoil, my stress, and my problems, and turn them into plans on healing and freeing myself from lifelong chains that had bound me. I felt calm and relaxed and capable of dealing with who I was, good or bad. (...) I am able to look deeper inside myself to make good, sound decisions based on my true beliefs and morals."*[66]

These reports show that some cannabis users feel that a high can help them to understand aspects of their personality better, to "look deeper" into their emotions, moods, bodily sensations, and character traits. In his study "On Being Stoned", Charles Tart also mentions that cannabis users feel that a high can help them to better introspectively access their current bodily sensations and feelings – which is, introspection in the more narrow philosophical sense. Many of the users surveyed by Tart confirmed the following effects as a "common" effect of a cannabis high:

"My skin feels exceptionally sensitive"

"Pain is more intense if I concentrate on it"

"My perception of how my body is shaped gets strange; the 'felt' shape or form does not correspond to its actual form (e.g. you may feel lopsided, or parts of your body feel heavy while others feel light"

"I feel a lot of pleasant warmth inside my body"

"I am much more aware of the beating of my heart"

"I become aware of breathing and can feel the breath flowing in and out of my throat as well as filling my lungs"[67]

In the following, I want to explore several different ways in which a cannabis high may enhance our introspection in both senses, in the more narrow, technical philosophical sense concerning feelings, thoughts, and sensations, as well as more broadly construed concerning moods and character traits.

The Body Mapping System and the Enhancement of Bodily Sensations

According to neuroscientists like A. D. Craig and Antonio Damasio, we all have an *interoceptive* sense which gives us a sense of the body's interior. Accordingly, this inner sense rests on a representational mapping system, which developed for us to observe our internal states such as

"(...) pain states, body temperature, flush, itch, tickle, shudder, visceral and genital sensations; the state of the smooth musculature in blood vessels and other viscera (...). "[68]

Our brain has a body mapping system that represents bodily parts. The hands, for instance, are represented in much larger areas in the brain than the areas that represent other body parts, because the brain needs more space to provide the highly complex and fine-tuned motor control functions of fingers and the hand itself. These regions in the brain grow, by the way, if you learn to play the violin for a long time because more fine-tuning motor control is needed and takes space.

In their groundbreaking book, *The Body Has a Mind of Its Own* authors Sandra and Matthew Blakeslee focus on recent research on this ingenious body mapping system – a system that has been underestimated so far in the cognitive neurosciences. They describe a system of 'flesh bound' somatic senses which fall into several categories, each subserved by different populations of receptor cells, such as the sense of touch, thermoception (feeling cold or hot), and nociception (detecting various kinds of pain, including piercing pain, heat pain, chemical pain, joint pain, tickle, and itch). Two other somatic senses would be proprioception, a sense of your body's position and motion in space, as well as the sense of balance. All these somatic senses deliver information to the brain's "body maps":

"Every point on your body, each internal organ, and every point in space out to the end of your fingertips, is mapped inside your brain. Your ability to sense, move, and act in the physical world arises from a rich network of flexible body maps distributed throughout the brain – maps that grow, shrink, and morph to suit your needs. "[69]

Could it be that cannabis has a systematic effect on this system, intensifying the signals and, thus, leading to a redirection of attention to perceive and feel our own body more intensely? We already know that the endocannabinoid system in humans has a variety of functions and regulates many cognitive as well as physiological processes, including processes having to do with bodily perception, such as pain sensation and the regulation of appetite. This knowledge to some degree confirms that cannabis and its phytocannabinoids may affect endocannabinoid receptors which are involved in regulating some somatic senses. Importantly, then, it could be highly interesting for endocannabinoid researchers to look at the reports of cannabis users to come to a better understanding of which processes in the endocannabinoid system may be further implicated.

Enhancement of "Reflective Contemplation"

As to the introspection of moods, more complex emotions, such as love, or the feeling of guilt, and character traits, it is obvious that this is not a kind of 'direct' inner observation. If I introspectively realize that I am a courageous person, I have to make a judgment involving the evaluation of many autobiographical memories, of "courageous" patterns in my behavior compared to the behavior of others, and involving my understanding of the concept of "courageous." We could also call this kind of introspection "reflective contemplation." A cannabis high might have several effects on cognition which could lead to the many enhancements described by users.

Let me briefly explain five effects of a cannabis high well known to cannabis users which I would like to highlight here: the hyperfocus of attention, the redirection of attention, enhanced episodic memory, enhanced imagination, as well as enhanced pattern recognition.

One of the most important acute effects of marijuana is hyperfocus in attention, an effect which I also like to call the *"Zen effect"* of marijuana because Zen tells you to concentrate on one thing or activity at a time. This attentional focus often leads to an intensified and more detailed experience of sensations and of a strong feeling of being in the here-and-now, but it can also lead to hyperfocus on a stream of thought or memories of past episodes in your life. Our attention seems often to become redirected, especially under large doses of cannabis, to our inner world, our episodic memories, our imagination, as well as to planning or analyzing remembered events or situations.

Myriad cannabis users have reported not only the hyperfocus of attention but also the enhancement of episodic memories. Users often vividly remember past events, events that they have often long forgotten, with incredible detail. Also, it is a very commonly reported effect of a cannabis high that users can imagine things better – and importantly, imagination does not only mean visualization, but it could also be auditory, tactile, taste, or olfactory imagination. We have seen that a pre-synesthetic effect can lead to enhanced pattern recognition, which we find over and over in reports of cannabis users. When we think of pattern recognition, we usually only think of the recognition of visual patterns, such as seeing a checked pattern on a pullover. We have to think of pattern recognition as a highly intellectual ability, though, which encompasses much more than that, such as recognizing an aggressive behavioral pattern in somebody's gestures during a conversation, or the recognition of a pattern in somebody's behavior that may indicate an attitude of pride.

Enhancement of "Reflective Contemplation"

Now, how could these five enhancements (hyperfocus of attention, redirection of attention, enhanced episodic memory, enhanced imagination, enhanced pattern recognition) affect our introspection? I think this is pretty easy to see. Let us assume you are reflecting on whether you tend to act naively romantic in your relationships, as your best friend just told you. A cannabis high can help to redirect and hyperfocus your attention on your episodic memories and your inner stream of thoughts. Nothing else matters now and you delve into associations of memories. At this point, enhanced pattern recognition is already happening, mostly at an unconscious level, because you have to take this notion and go hunting for similarities of patterns of "naively romantic" behavior in your life, hundreds of episodes that could span from your early adolescence to decades later. During a high, you better see associations and similarities between episodes and behaviors you would normally not see. Now you can connect the dots better.

Note how your enhanced ability for imagination might also play a crucial role in the success of reflective contemplation concerning your character trait: if you want to judge whether you are a naively romantic person in general, you do not only think about your past, but you may try to imagine future scenarios in which you would have to make decisions in a relationship. Would you trust your beloved one again in a relationship even if she confessed to having betrayed all of her or his ex-partners several times?

During a cannabis high, you can often imagine situations like these more vividly and reflect on how it would be for you, what you would feel, and how you would act. Thus, an enhanced capacity for imagination could generally help you to come to valuable insights about your dispositions and character traits, such as the insights expressed in this statement by a college student:

"Pot is very therapeutic to me. When I'm stoned, I can see myself. I can list my strengths and my weaknesses, and goals. My mind is clear and eager to learn and understand, even when I have to understand awkward things, like those parts of my personality that I don't want to change. I can see parts of myself that I don't like, without hating myself in the process. I've learned things about myself that I have brought into my life when I haven't been stoned, such as how to be less self-centered, and how to be more low-keyed about myself, and less anxious in the presence of others."[70]

Clearly, then, we can see how many of the effects of a cannabis high on our mind could play an important role in the enhancement of our ability to reflectively contemplate and to come up with deep insights about ourselves under favorable conditions.

An Amazing Potential

Let me stress again that my claim is not that cannabis automatically enhances your introspection. My claim is that cannabis has the potential to enhance your introspection – but to use this potential you have to use it appropriately, with the right dose, the right attitude, an apt variety of cannabis, with the right set and setting, and with knowledge and skills to "ride a high."

An enhancement of our introspection can also help us

get in touch with our feelings and other subjective states to help us understand others as well. Another anonymous contributor to Grinspoon's website *marijuana-uses.com* reports:

> "*Friendships and Relationships, especially those involving sexual and romantic intimacy, can be developed and deepened by the use of marijuana with others. Marijuana tends to cause introspection and by altering one's habits of thought, yields new perspectives on who one is and how one works, psychologically. Hence, marijuana works as an effective catalyst for understanding oneself and others, and discussing and developing one's relationships with other people.*"[71]

Evidently, then, enhancing our ability to understand ourselves, can also lead to a profound new understanding of others. Cannabis, then, has the potential not only for the enhancement of introspection but also for empathic understanding, as we will further discuss in the next chapter, as well as for personal development.

It is about time that we recognize that many people use cannabis not only for "recreational purposes", but are using it as a tool to find out who they are, to understand themselves and their relation to others better, and to grow as a person. If we think of it this way, then, "recreational use" seems shallow and not quite satisfactory to cover the variety of enhancement uses of many cannabis consumers. Yet, from a linguistic point of view, it is not that far off. Just insert a hyphen and start calling it the "re-creational use" of cannabis.

Epilog

There is a general problem, of course, when it comes to the question of how a cannabis high or any other altered state of consciousness affects our introspective abilities. If our approach includes a psychonautic or phenomenological component, which relies to a certain extent on our introspection, our very research method is in question, as a distorted sense of perception during an altered state of mind may provide systematically misleading "data" on this very state of consciousness. Importantly, this also holds for anecdotal reports of others and any study designs involving humans and their self-reports of their conscious states during an altered state of consciousness, as their reports will of course depend on their introspective abilities.

Exploring this topic would be a lovely theme for a dissertation thesis, but let me only make a few comments here. First, we do not have to merely rely on introspection, as our mental states can also to some degree be judged by our behavior. So, for instance, if I introspectively come to believe that I am especially creative during a high, but really, my friends and I later see that everything I produce during a high is below the standard of what I would normally produce, we already have a good corrective to my introspective judgment from an interpersonal evaluation.

Philosophers today know that introspection is not infallible. I may be wrong when I come to introspect my feelings and come to believe that I am in love with my neighbor Amy – maybe my psychoanalyst will sort it out for me and we later come to the conclusion that I am sexually attracted to her, but that my feelings are not really what we would all agree on as being love. Still, most philosophers would agree that our introspection grants a conscious subject a certain *privileged access* to his inner mental realm, which makes it an indispensable source of knowledge when it comes to exploring conscious states, including altered states of consciousness.

Are we all at the same level of ability when it comes to introspection? Certainly not. Consider the case of a wine connoisseur. A wine connoisseur will not only know about different grapes, their ingredients, as well as fermentation, and other processes that shape a great wine. They will also have to learn how to name and discriminate fine nuances in a very complex taste experience of sipping a glass of wine. Briefly, then, introspective skills can be honed, and they also involve a deep knowledge of objects which affect conscious states, a knowledge that will also find its expression in a refined language. Here is a short passage from an article about wine slang that nicely illustrates this point:

> *"You slink into the classy restaurant and take your seat, perusing the lengthy drinks menu before ultimately giving up and asking the man in a sharp suit for a recommendation. The next thing that comes out of his mouth seems like a spy code:*

> *'We have a great vintage of Cabernet Sauvignon from a winery on the Wahluke Slope in the Columbia Valley AVA. Lean, restrained, acidy, and alive, with puckery tannins and a velvety mouthfeel. It was the talk of the Grand Tasting at last year's Taste Washington.'"*[72]

Note, however, that wine sommeliers usually focus on a description of the complex taste experiences of wine and not on its manifold effects on consciousness, such as its ability to focus one's attention or its effects on memory or pattern recognition. If we want to get better introspective reports on the cannabis high as it affects our mind, researchers would benefit from cannabis users who understand the cognitive lingo and are, thus, better equipped to come to interesting introspective observations.

Also, naturally, it would be interesting for them to be proficient in various techniques of mindfulness, as they will help them in self-monitoring. I am hoping that my work can contribute to educating a new generation of Cannaficionados to take our research of the cannabis high to the next level.

(14) A Silver Haze flower shortly before harvest. The white stigmas turn increasingly reddish-brown as they mature, indicating that its THC decays to CBN (cannabinol), which is believed by some experts to cause sedation and drowsiness, probably even more so in conjunction with the degradation products of other chemical compounds in the plant. Still, CBN, potentially has medicinal benefits, for instance to probably bring relief for muscle pain.

VI. Cannabis, Mirror Neurons, and Empathic Understanding

"First of all, if you learn a simple trick, Scout, you'll get along a lot better with all kinds of folks. You never really understand a person until you consider things from his point of view . . . until you climb into his skin and walk around in it."
Atticus, in *To Kill A Mockingbird* by Harper Lee

What is empathic understanding? Let me first clarify this question with an initial definition to avoid misunderstandings that may arise out of our everyday use of the term "empathy". Many people seem to equate empathy and empathic understanding with what specialists would now call *affective empathy,* the ability to *perceive and share* emotions or moods with others. Those who score high on affective empathy are intensely "feeling the pain of others" in themselves. In other words, they feel "com-passion" for others; they suffer with them. On the other hand, *cognitive empathy* denotes our ability to understand others, to describe, explain and predict how they think, feel, and behave. On the face of it, it seems that there are cases where we can *cognitively empathically* understand somebody without going through his or her mental state. An experienced doctor sees your bone fracture and she will usually be able to understand you are going through a painful experience. She can explain and predict some of your behaviors and feelings, but she does not have to go through those feelings herself.

After seeing a lot of suffering, most doctors have learned to *emotionally regulate* their feelings to some degree so that they do not feel the full impact of the amount of suffering of others.[73] Too much compassion could potentially lead to early burnout and would neither be helpful to themselves nor to their patients. In this sense, as we have grown up, most of us have become "experienced doctors." We can emotionally regulate our feelings in the face of vast human and animal suffering around us in this world.

Without a doubt, the ability to empathically understand others is one of the most advanced cognitive skills in animals, evolutionarily speaking. We see highly developed empathic behavior in rats, monkeys, dogs, elephants, and many other animals. Empathic understanding is not a skill that only evolved in humans. Yet, arguably, the rapid evolution of empathic skills allowed the human species to mimic and learn, collaborate, and build cultures that led to our evolutionary dominance.

People lacking empathic skills can have a radical disadvantage in society. As Harvard psychologist Daniel Goleman puts it,

> *"If your emotional abilities aren't in hand, if you don't have self-awareness, if you are not able to manage your distressing emotions, if you can't have empathy and have effective relationships, then no matter how smart you are, you are not going to get very far."*

Goleman makes a strong point, but let me comment on his claim that without empathy, *"we are not going to get very far."*

Sadly, our society today seems to be still dominated in large parts by psychopaths and sociopaths, as it has probably been from the beginning of our evolutionary appearance. I am writing these lines a few days after Russian dictator Vladimir Putin waged war against Ukraine. Goleman is right in the sense that a sociopath like Putin will never lead a happy and fulfilled life like somebody with a true capacity for empathy and compassion. In this sense, he will not *get very far*. But we have to understand that personalities with a severe lack of affective empathic understanding can be very effective and get very far in a different sense: they can still "coldly" empathically understand and predict others and their suffering. They know, for instance, how to instill fears. They lack or can suppress feelings of compassion and can scrupulously exploit and manipulate others by using sophisticated mechanisms for instilling fear. Sociopaths like Trump are often very powerful, also, because they know well how to exploit the compassion of their followers and by manipulating them into believing, for instance, that their leader is a victim himself.

As humans evolved they learned to empathically understand other members of their species, but also, other animals. This highly evolved empathic understanding, however, never meant that humans were purely compassionate beings: we have strong evidence that already tens of thousands of years ago, the human species, eradicated almost half of the animal species in the regions they migrated to. Their empathic understanding of animals and their fellow empathic communication amongst each enabled them to collaborate by the dozens to hunt, encircle, and kill a whole herd of wild horses.[74]

Let us keep this in mind, then, in our discussion of cannabis as potentially temporarily enhancing empathic understanding. The often reported enhancement of our ability to empathically understand others during a high can be incredibly valuable and useful, but it is yet another question if this effect could make a difference for societies as a whole, as more and more people in cultures around the world could be using cannabis soon. It is an interesting question, therefore, whether cannabis can enhance not only cognitive empathic understanding but may also enhance affective empathic understanding.

Reports about the Enhancement of Empathic Understanding during a Cannabis High

Cannabis users have reported various ways in which their empathic understanding becomes enhanced during a high. The psychologist Charles Tart's study *On Being Stoned* (1971) names the following two descriptions of effects confirmed by many students who participated in the survey as "*characteristic*" of the cannabis high:

> "*I have feelings of deep insight into other people, how they tick, what their games are, when stoned (...)*".

> "*I empathize tremendously with others; I feel what they feel; I have a tremendous intuitive understanding of what they are feeling.*"[75]

For moderate levels of a high, Tart also found that fairly frequently consumers would agree with the following statement:

> "*I feel so aware of what people are thinking that it must be telepathy, mind reading, rather than just being more sensitive to the subtle cues in behavior.*" [76]

Here is a report from a cannabis user cited in another book, *High Culture: Marijuana in the Lives of Americans*, by William Novak, which also mentions "mental telepathy":

"When I'm stoned with a very good friend, we just sit there and watch messages bounce back and forth between us, like neutrons. It happens rapidly, and we can feel it in an almost physical way. I often get onto a higher plane of communication with good friends when we smoke together. It almost seems as if we are experiencing mental telepathy, with communication going on rapidly (...)" [77]

Another report in Novak's book gives us a hunch on how cannabis might lead to the enhancement of empathic understanding during a high:

"After a joint or two, I find myself paying more attention to what the other person is saying, rather than hearing only the words he uses in trying to get his point across. By keeping track of his mannerisms and his tone of voice in a more concentrated way than usual, I can more fully understand his point, and can respond more directly than normal." [78]

This user report mentions a shift of attention to "his mannerism, and his tone of voice". We know from other reports that a cannabis high can lead to hyper-focusing and redirection of attention. A shift of our attention to a stronger focus on body language and the "tone" of a statement certainly may help to understand others better. This is why dogs understand us so well, why they feel our sadness, joy, or pain, without being able to understand our language. To a surprising extent, they attend to and understand our body language, gestures, and the tone of our voice, and they smell our fear or excitement. Note, also, that to read a "tone of voice" or a certain mannerism, you have to be able to decipher a complicated pattern. We have numerous reports from marijuana users on all kinds of occasions that they can recognize patterns during a high that they have not perceived before; patterns in music, art, nature, or a behavior.

Hyper focusing, a change of the direction of attention to body language, and enhanced pattern recognition are all effects that have been independently and in detail reported by cannabis users, and the report above well describes how these effects could in combination lead to an enhancement of our empathic understanding.

Also, cannabis users have consistently reported vividly remembering past events during a high. They can retrieve memories that seemed to be long gone, and feel like they "re-live" them. The following report of an anonymous 19-year old computer programmer about his high experiences beautifully illustrates this:

"Memories seemed to force themselves upon me, very rapid but very gentle. I started to remember things in my childhood that made me truly happy and joyful. Things I had either forgotten or just simply didn't give the time of day to. I remembered raising my hands up as a signal for my mother that I wanted to be carried and the utter joy I felt when she would reach down and pull me up to her chest. I realized how much she really did, in fact, love me when I remembered how I longed for her goodnight kisses, of which never ran dry. I remembered the very simple joys of my very simple existence and marijuana helped me relive them all over again."[79]

It is easy to see how an enhanced ability for episodic memory retrieval during a high can help to understand other people better empathically. If you vividly remember episodes of your feelings of *teen angst* during your last years of high school, you will probably understand your 15-year-old son better in a similar situation. If you have better access to detailed memories of how you felt during the day of your wedding, you will better understand the reactions of the young couple you are seeing at their wedding, pale, stressed, and yet happy, clutching their hands and smiling nervously.

These cognitive skills – episodic memory retrieval, attentional focus, the re-direction of attention to various patterns such as body language, and enhanced pattern recognition on various levels – are some of the crucial cognitive abilities for empathic understanding as well as for introspection, and many independent users' reports from various centuries suggest that all these can at some point become enhanced during a high.

The cognitive skills named above are of course not the only mental abilities important to empathic understanding. One of the most decisive cognitive skills for our ability to read the minds of others is our ability to imaginatively "slip into their moccasins", as the native American Indians say, or to "climb into someone's skin", as Atticus advised in Harper Lee's book, *To Kill a Mockingbird*.

For more than thirty years now, so-called "simulation theorists" in the philosophy of mind and the cognitive sciences have argued that our capacity to empathically understand other people, to predict, explain and describe their feelings, thinking, moods, and behavior, is crucially based in a special capacity to *imaginatively simulate* others in their situation, to imagine being in their skin and to feel their feelings, to think their thoughts.[80] We know many reports from cannabis users that they often feel that their ability to generally imagine situations becomes enhanced; they can imagine more vividly and in more detail. So, it makes sense that we take another look at the role of imagination and mental simulation based on it for modern theories of empathic understanding.

The Mirror Neuron System, Imitation, and Empathy

The simulation theory received new supporting empirical evidence when Italian neuroscientists found a class of neurons in monkeys that seems to be responsible for mimicking some behavior that a subject perceives in another. The revolutionary discovery of 'mirror neurons' began in the early 1990s when Italian neuroscientists accidentally stumbled upon what seemed to be a weird phenomenon in experiments with macaque monkeys. Giacomo Rizzolatti and his research team at the University of Parma in Italy had originally only intended to study motor neurons in the frontal cortex – neurons known to be responsible for our motor control system. With tiny electrodes attached to individual cells in a monkey's brain, they wanted to find out how certain hand-grabbing movements are initiated in the brain. These motor neurons consistently showed the expected activity when a monkey moved his arm and picked up a peanut.

Surprisingly, however, the Italian research team witnessed how those monkey motor neurons also fired when the monkey just watched a lab assistant picking up the peanut. At first, the scientists could not believe what they saw: the monkey did not move at all, so why did the motor neurons fire? Was there something wrong with the cable connections or their measuring instruments? More tests, however, showed that there was nothing wrong with the wiring or the measuring system. These monkeys' motor neurons for controlling their hand "mirrored" the hand-moving activity that the monkey perceived in somebody else. They mirrored the perceived activity, even when the monkey only *heard* an activity without performing the act itself.

We all know about the processes of mirroring feelings and sometimes even the behavior of others from our own experience. When you observe somebody in pain, you tend to mirror his feelings to some extent and, often, to some degree even mirror his behavior, such as his cringing facial expression of pain. Yawning is so contagious to others that if somebody yawns in a classroom, many others will follow and mimic the behavior and "mirror the action." I bet that you feel like yawning now even only reading this. In toddlers, this unsuppressed motor mimicry can be observed up to the age of about two and a half years. If a toddler sees another toddler cry, he will often cry himself, too. The toddler sees the crying behavior of the other and mimics not only the behavior, but feels the feeling of the other as if it was his feeling. He is in pain himself. In other words, the toddler has a lot of effective empathic understanding, but he has not built up his ability to regulate his empathic feelings.

From around 2002 onward, the interest of neuroscientists in human mirror neurons became a boom that lasted until 2013. Many neuroscientists were now deeply convinced that the mirror neuron system was crucially involved in cognitive motor control, mimicry, and imitation behavior, in the processing and understanding of language, introspection, and empathy.[81] Mirror neurons became the new big thing in neuroscience. Neuroscientist V.S. Ramachandran expressed his highly optimistic expectations about mirror neurons already in 2004:

> *"The discovery of mirror neurons in the frontal lobes of monkeys, and their potential relevance to human brain evolution – which I speculate on in this essay – is the single most important "unreported" (or at least, unpublicized) story of the decade. I predict that*

mirror neurons will do for psychology what DNA did for biology: they will provide a unifying framework and help explain a host of mental abilities that have hitherto remained mysterious and inaccessible to experiments. (...)

I would argue (...) that mirror neurons are Necessary but not sufficient: their emergence and further development in hominids was a decisive step. The reason is that once you have a certain minimum amount of "imitation learning" and "culture" in place, this culture can, in turn, exert the selection pressure for developing those additional mental traits that make us human. And once this starts happening you have set in motion the auto-catalytic process that culminated in modern human consciousness. "[82]

Ramachandran also hypothesized that autism could be related to a malfunctional mirror neuron system, a thesis that became known as the *"broken mirror hypothesis"* of autism.[83]

Around 2014, however, mirror neuron research began to decline. Newer studies showed a more complicated picture that did not quite support the grandiose expectations of researchers in this emerging field. Especially, the broken mirror hypothesis became heavily criticized, with mixed interpretations of existing studies. Still, in their newer overview article published in 2021, "What happened to Mirror Neurons?", Cecilia Heyes and Caroline Catmur summarize the current state of the evidence as it concerns the role of human mirror neurons in various cognitive functions and attests that further research is still promising:

"For action understanding, multivoxel pattern analysis, patient studies, and brain stimulation suggest that mirror-neuron brain areas contribute to low-level

processing of observed actions (e.g., distinguishing
types of grip) but not to high-level action
interpretation (e.g., inferring actors' intentions). In the
area of speech perception, (...), there is compelling
evidence for the involvement of the motor system in
the discrimination of speech in perceptually noisy
conditions. For imitation, there is strong evidence from
patient, brain-stimulation, and brain-imaging studies
that mirror-neuron brain areas play a causal role in
copying of body movement topography. In the area
of autism, studies using behavioral and neurological
measures have tried and failed to find evidence
supporting the 'broken-mirror theory' of autism. "[84]

Note that the lack of empirical evidence for the
broken mirror theory about autism could have to do with
the complexity of autistic spectrum disorders.[85] Most
importantly, it is now known that it is a myth that those
diagnosed with autism spectrum disorder (ASD) are
generally not very emotional or unable to empathically
perceive emotions in others. According to newer studies,
approximately 50% of those diagnosed with ASD have
alexithymia, a condition characterized by difficulties in
understanding and describing one's own emotions. This
explains why many parents have observed that contrary
to the stereotypical image of ASD, their children showed
typically, or sometimes even excessive empathy.[86] We will
come back to these insights when we discuss the potential
therapeutic effects of cannabis on children with ASD
reported by many parents.

(15) A flower of a The Pure variety. On the macro level, we can see how different cannabis varieties can be. Today, we cannot necessarily tell exactly from plant morphology what the genetic ancestry is and what kind of profile the plant has in terms of its distinctive production of cannabinoids, terpenes, and flavonoids. But we can still see that plants can be very different, visualizing the fact that the medical effects and the effects on the mind can vary greatly between different cannabis varieties.

Endocannabinoids, Phytocannabinoids, and The Mirror Neuron System

Could it be that this mirror neuron system or some other related specialized neuronal system that subserves functions of empathic understanding becomes enhanced during a cannabis high? Such an enhancement could lead to the subjective feeling that we can more easily slip into the moccasins of others, that we can feel what they feel, that we almost slip into the skin of another person. Many users have noted a better ability to imagine situations in general, visually as well as auditorily or in other sensory categories; but we also have many very specific reports from users about how they were better able to imaginatively put themselves in the shoes of other people, to "slip into their moccasins".

Here is one of many beautifully detailed reports from cannabis users:

"Whilst stoned, I found it easier to put myself in the place of others. I could understand how people might believe any number of seemingly "irrational" or dense, impenetrable ideas. Marijuana opened me up to the existence of so many different views of the world, views I need not share to fathom or empathize with. (...) Let me give you one recent example. (...) One night, amid a marijuana-induced reverie, I got to thinking about the real person we call Jesus. These days, we think of him as some ethereal figure in some far-off land shrouded in historical mists of time. (...) I found myself imagining that at one time in history, this Jesus character was a living, breathing human like myself. Intellectually, I already knew this, but I increasingly felt as though I might be capable of fathoming what his disciples and apostles felt. His followers were in his presence, they looked into his eyes and heard his words, and believed they were looking at God. It was only in this state of consciousness that I could truly imagine what it might have been like to be in the presence of this man and truly believe, and by extension, possibly experience what so many people on Earth experience during moments of great religious feeling and devotion. (...) In this case, being stoned allowed my mind to

circumvent its ordinary non-religious bent, and if only for a few moments, come to know what the truly religious feel. While I don't subscribe to the tenets of Christianity proper, I have come to understand how real it can all seem for people (...). "[87]

This report is interesting, also, because it shows how empathic understanding plays a role even if nobody else is in the room; the empathic understanding here proceeds only via a re-interpretation of memories.

Cannabis and Autistic Spectrum Disorder

If a cannabis high can enhance empathic understanding, could it also help those diagnosed with autistic spectrum syndrome – or at least a subgroup of those who are having problems with their empathetic understanding?

Autism spectrum disorder is a pervasive developmental disorder. Children diagnosed with this mysterious syndrome often show problems in developing human relationships, some have abnormal speech or cannot speak at all, and often, have problems with social, emotional, and other communication skills. Some of them are often engaged in almost ritualistic repetitive behaviors instead of imaginative play, and some do seem to have special problems understanding the feelings and needs of others. Some children with ASD seem to be very aggressive, while others are not. Generally, the diagnosis of "ASD" seems to encompass a very heterogeneous group of people with very different symptoms. While it has become clear in recent years that not all of those diagnosed with ASD show similar deficiencies in empathic understanding abilities, it is well worth looking at some of the reports of parents of autistic children the many ways in which cannabis was therapeutic for their kids.

Around a decade ago, stories of a few mothers surfaced in the news in the U.S., mothers who reported they had given medical cannabis to their kids and thereby radically improved their health condition. When I heard about these stories I was of course curious to find out exactly how cannabis helped those kids. At first, I was somehow disappointed when I read mostly details about how cannabis seemed to help calm children with ASD, make them happier, how it helped them to lose some tics, or to regain appetite. All this was wonderful for these kids and their families, of course, but it was not exactly what I was looking for. I was fascinated, then, to read descriptions about cannabis helping kids diagnosed with ASD to some degree "free" them from their repetitive behaviors, which reminded me a lot of the use of Indian sadhus, who have always used the cannabis high to free themselves from the repetitiveness and the routines of our thinking. But I somehow knew there would be more – and I found it. I was amazed to read the story of Marie Myung-Ok Lee, a Korean-American mother author, and essayist, who wonderfully described the following change in her son "J", a 9-year old boy who – despite being on several heavy prescription drugs showed from 50 up to 300 violent aggressions in a school day. After the removal of almost all other medical drugs, advised by Lester Grinspoon (which I only found out much later), and medicating "J" with a specific cannabis variety, her son showed much fewer aggressive behaviors, sometimes no aggressive behaviors for a day or two.

Marie Myung-Ok Lee reported truly amazing behavioral changes in her son during medication:

"The big test, so far, has been a visit from Grandma. The last time she came, over Christmas, J hit her during a tantrum. This time, we gave him his tea, mixing it with goji berries to mask any odor, although it occurs to me that my mother, a Korean immigrant, probably doesn't even know what pot smells like (and it smells a lot like ssuk, a Korean medicinal herb). She remarked that J seems calmer. As we were preparing for a trip to the park, J disappeared, and we wondered if he was going to throw one of his tantrums. Instead, he returned with Grandma's shoes, laying them in front of her, even carefully adjusting them so that they were parallel and easy to step into. He looked into her face, and smiled." [88]

J's behavior shows a remarkable act of sophisticated empathic understanding, given the severeness of his specific autistic condition; it required some understanding of his grandmother's needs and feelings to neatly arrange the shoes like this, and his smile must have been priceless for her. His mother mentioned another extremely important detail about his medication:

"J. still can get overexcited if he likes a food too much, so sometimes when he's eating, my husband and I leave the room to minimize distractions. The other day, we dared to experiment with doenjang, a fermented tofu soup that he used to love as a baby. The last time we tried it, a year ago, he'd frisbeed the bowl against a tile wall. (...) We left J. in the kitchen with his steamy bowl and went to the adjoining room. We waited. We heard the spoon ding against the bowl. Satisfied slurpy noises. Then a strange noise that we couldn't identify. A chkka chkka chkkka bsssshhht doinnng! We returned to the kitchen, half expecting to see the walls painted with doenjang. Everything was clean. The bowl and

spoon, however, were gone. J. had taken his dishes to the sink, rinsed them, and put them in the dishwasher – something we'd never shown him how to do, though he must have watched us do it a million times. In four months, he'd gone from a boy we couldn't feed to a boy who could feed himself and clean up after." [89]

This activity may sound trivial to us normal people, but the fact that "J" was able to engage in that kind of mimicry behavior and clean up the dishes during a high was astounding to his mom, who knew him well and saw how much of a leap in cognitive function this was.

Mirror Neurons and the Endocannabinoid System

As I have laid out above, the findings of the mirror neuron system and its role in empathic understanding and ASD are still being debated in the scientific community. Still, I think that the reports of cannabis users should give us enough reason for an initial suspicion, a hypothesis that needs to be taken to the laboratory with brain imaging techniques. Can a cannabis high temporarily affect human empathic understanding, maybe by having effects on our mirror neuron system? Or does cannabis help in some other way to stimulate other neuronal systems subserving our skills to empathically understand others? And if a cannabis high can have those effects, are they mediated by an endocannabinoid system that has some functional connection with those neuronal systems that subserve empathic understanding? Or could the endocannabinoid system itself play a direct functional role in some of the cognitive skills needed for empathic understanding? How do other phytocannabinoids like CBD or the terpenes affect our ability to empathically understand, and how are those effects mediated? Can CBD help with autistic spectrum disorder? According to recent findings,[90] we can assume that THC probably works better, but CBD may be a promising therapeutic option for some subgroups of those diagnosed with autistic spectrum disorders.

If the endocannabinoid system does indeed play a fundamental role in our empathic understanding, what would the implications be for the treatment of autism or other disorders, for the use of cannabis in psychotherapy, or the personal use of cannabis?

These questions are urgent, and the answers could have enormous importance for millions of people – patients and responsible adult users alike.

If a cannabis high can indeed help us temporarily enhance empathic understanding during a high, then it could be of tremendous help for all of us; to understand our kids better, to lead better relationships with our partners, to extend our range of understanding to very different people, to collaborate better with our colleagues and to excel at work, to gain insights in patients as psychotherapists. Again, note that as in the case of other mental enhancements, even if cannabis has the potential to help us with empathic understanding, we would have to learn how to use cannabis as a tool for this ability to avoid cancelation effects.

Is the enhancement of our empathic understanding during a high more on the cognitive or more on the affective side? I think it should be clear by now that the cannabis high seems to enhance both the cognitive part of empathic understanding, as it can for instance help us to see new patterns in behavior. But all in all, there seems to be a very strong effect on affective empathy; we have strong introspective access to our feelings and memories as we simulate others and slip into their moccasins; we emotionally connect to others by taking their perspectives and feeling their feelings. For some cannabis users, this can be quite a life-changing experience.

A Solipsist, Reaching Out

A few months after I published another essay on cannabis and the enhancement of empathic understanding for an expert blog of a Dutch cannabis seed bank in 2015, a 20-year old student from New York University (NYU) contacted me. In his first email to me, he wrote:

> *"I have struggled with empathy as a kid. I think it never was taught to me because of complex traumas in my family, and as a result, I grew completely unable to realize other people were separate from me. For the longest time, the world was simply an extension of myself, my own video game. But cannabis brought me back to reality. Other people do exist. They are separate and have their thoughts, and emotions and some of them suffer like I do. I have smoked and started thinking about how my girlfriend/mother/ brother must have felt for the first time, and it allowed me to feel my sadness with much more deepness and openness than usual. Smoking has given me insight into my existence and into empathy like never before. Like you say, better pattern recognition and a newfound capacity to imagine what it is to be in another person's shoes, to know that they actually exist."*

My curiosity was immediately sparked, of course, and soon after we had a long skype conversation about his story and how he felt. Of course, I needed to know what he meant by *"the world was simply an extension of myself"*. It turned out he had truly felt and believed in a view that philosophers would attribute to the position of a *metaphysical solipsist*; the view that only the own self exists. Everything else is a dream, an illusion, and only one's own consciousness exists. There is no other consciousness or outside world.

In our conversation, this student told me some fascinating details about his use of cannabis. He said when he was under the influence of cannabis, he would feel connected and knew there were other people, but he would gradually lose this feeling as time passed and the effects of cannabis wore off. This feeling of connectedness would fade away. A few days after his cannabis high he would still remember and know that there are other people out there in sober condition, but he would not feel it anymore. He felt alone again. And he wanted to belong.

Could it be that the endocannabinoid system is involved in the underlying neural mechanisms which provide this profound and fundamental feeling of connectedness? It could well be that a cannabis high enhances certain cognitive functions like episodic memory retrieval, imagination, and pattern recognition. Each of these enhancements could explain why so many cannabis users have felt they can better empathically understand others during a high. Yet, given the detailed reports of cannabis users including the story of the "solipsist", there seems to be something even more profound going on. Research in this direction could help us to find treatment options not only for autism or mental syndromes that bring about problems in empathic understanding. It could also help us to come to a better understanding of how humans generally manage to understand others, and why we do not feel alone in this world.

(16) The mature flower of a Mexican Sativa developed on the basis of a variety that originated in the southern province of Oaxaca in Mexico.

VII. Cannabis, Love, Sex, and Tantra[91]

"If one wanted to depict the whole thing graphically, every episode, with its climax, would require a three-dimensional, or, rather, no model: every experience is unrepeatable. What makes lovemaking and reading resemble each other most is that within both of them times and spaces open, different from measurable time and space."

Italo Calvino (1923-1985),
If on a Winter's Night a Traveler

Can a cannabis high enhance your love life, if skillfully used under favorable circumstances? My short answer is: yes, in many ways, and many more ways as are usually cited in modern columns about this subject. The cannabis high is a multidimensional altered state of consciousness characterized by a whole range of altered cognitive functions, and many of them can play a role in our love lives.

A cannabis high and sex can go incredibly well together. Many experienced cannabis users appreciate the aphrodisiac effects of various cannabis varieties, but usually understate the wide spectrum of psychological enhancements of a cannabis high. Many of the cannabis-induced mind enhancements I have described in my work can positively influence sexual experiences, such as the intensification of the experience of touches, kisses, or orgasm, a strong feeling of awe and to be in the here-and-now, relaxation, the enhancement of body image perception, or the subjective slowdown of time. But this is only the beginning.

The Multidimensional High and the Sexual Experience

Many modern cannabis users have reported in detail how a cannabis high enhanced their love lives. Here is a fragment of a beautiful letter of an American husband, who explains his cannabis use to his wife, also because she was anxious about his use of what she perceived as a dangerous illegal drug:

> *"Speaking of sexual relations, remember all of those times when you have told me, "Wow, I have never felt like that before." Where, at the end of our lovemaking, a simple touch would leave you trembling on the bed, shaking and laughing because it all felt so good? Where two hours of foreplay was followed by hour-long love-making sessions that left us holding each other tightly and remembering the love that we have for each other? That was, oddly enough, a product of not only my love for you but also pot. Don't get me wrong, the sex is great without it, but when I'm under the influence of this "horrible" substance, I slow down and enjoy each moment, making particular sure to give attention to everything you need and completely satisfy you."*[92]

Lovemaking is more than sex, and this letter gives us a wonderful description of some of the aspects that make a difference: *"making sure to give attention to everything you need and completely satisfy you."* Under the influence of marijuana, this husband became more empathic and better able to feel and understand the needs and desires of his wife.

Innumerable other modern cannabis users have rediscovered various uses of the cannabis high to enhance their sex lives. Here is how I would summarize their reports:

During a high, you focus your attention on the *here-and-now*; you do not think about a previous verbal fight with your partner, or about an upcoming exam the next day. You relax, take your time. The pressure and the stress are gone. In this space, you can better concentrate on your own body, on your own needs and desires, as well as on your partner, his or her current needs, longings, and desires. For some, this helps to arouse the desire for sex even before it starts. Your desire for sex becomes intense. During sex, bodily sensations are intensified and experienced in more depth; the touch of lips during kissing, scents, stroking – everything appears more vivid, but also more detailed in perception.[93] You are not distracted, you stay in the here-and-now of experience. Often, you awe at the freshness of experience – a kiss feels as if you have never kissed before, like the first kiss of your adolescence. You seem to be separated from the past and future, floating. Time has slowed down. You experience everything in slow motion. As the high becomes intense, time does not seem to exist anymore. As the cannabis user Carl Sagan once remarked, this perceptual slowdown of time leads to a prolonged sensation of orgasm, and, I would add, a prolonged sensation of the whole experience of lovemaking.[94]

Many of the effects described above have been quoted by others when it comes to the positive effects of the high on sex – but these are only the beginning. The enhancement of imagination, one of the most important, yet often ignored enhancements during a high, is described in a tale from the Arabian story collection *One Thousand and One Nights*, where a stoner obviously "trips" on a large amount of hashish and comes back to his senses in a public bath:

"He opened his eyes and found himself lying on the marge of the cold-water tank, amongst a crowd of people all laughing at him; for his prickle was at the point and the napkin had slipped from his middle. So he knew that all this was but a confusion of dreams and an illusion of Hashish (...) "[95]

During a strong high under such a large dosage, cannabis users have often reported "visual trips"; but even for moderate dosages of marijuana, imaginative abilities can become enhanced. This effect can be very beneficial, of course, but in large doses, the high can also lead to an introspective state of strong visual tripping which makes it hard to interact with a partner during a sexual experience.

Furthermore, as mentioned before, many cannabis users have described an enhanced ability for pattern recognition during a high. During high sex, enhanced pattern recognition can become helpful in many ways. It can help you to better recognize an emotion like sexual arousal expressed in your partner's body language, to recognize better how she responds to your actions, or help you to recognize a boring sexual routine you both have been going through for a while with one's partner – which may then help you to transcend the routine.

Many cannabis users gave us accounts of how they became too anxious, self-centered, and introspective during sex, which definitely did not help with their sexual experience. Dosage, skills, mood, attitude, and context obviously play a crucial role. Other, more experienced users who are better able to arrive at their sweet spot with a high, feel the high not only generates strong feelings of intimacy, but leads to a truly deeper empathic understanding of the other's character, needs, and wishes, based on various patterns in a partner's body language, in the tone of what is said, or in facial gestures. This enhancement of empathic understanding is fundamental. Sex is like a dance, a form of communication in constant need of new common explorations and fascinating discoveries.

High lovers often break with routines and become creative during their lovemaking. They become creative and use their imagination to refresh their love lives. Additionally, strongly depending on the set and setting, the high can bring an *anxiolytic* effect at certain doses, which helps to overcome moral and other inhibitions. This can be a major factor for entering a true energetic sexual flow.

Beyond temporary enhancements of the multidimensional sexual experience, cannabis can have a profound positive influence on us in a relationship and deepen our intimacy, as the following very perceptive report of a cannabis user illustrates:

> *"My wife decided to try it. Marijuana-induced conversations, the likes of which we'd never had in all our years of marriage, followed. I confessed that, far from being a novice with the sacred herb, I had eaten ganja brownies at concerts as far back as high school and college and loved every bite. (...) We've had good laughs about it, real laughs, for the first time since we've known each other.*
>
> *We learned that each of us had experienced deep intimacy with other people before we'd met, something we'd each long denied and always avoided discussing. We found that we both wanted to go for a romp on a hedonistic nude beach, and we discovered that neither*

*of us had ever found much satisfaction in the bigness
and richness of our previous life. We both confessed
to problems and dysfunctions in our families and
a lack of good marital role models. Interestingly,
none of this hurt. Marijuana, instead of leading us to
confrontations about hiding secrets or lying all these
years, filled us with a desire to discuss these things
all the more. (...) We saw each other, for the first time,
as flawed as everyone else and found in that a kind
of relief that dropped the weight of stainlessness and
conformity forever. (...)*

*Marijuana has landed us solidly into a groove of
change; it's broken down barriers between us, and
probably saved our marriage and family. We have a
heightened sense of what we're all about as a couple,
we're better parents, and more compassionate people,
and now know that at our core is a need for plain talk
and self-content. Most important, we're no longer
bluffing our way through life."*[96]

Obviously, all these effects described in this report
can help to create the basis for an entirely new sexual
relationship, too. Take alone the observation that cannabis
helped the couple to create an intimate realm with no
more room for bluffing. The lead singer of *The Doors* Jim
Morrison once said:

*"Sex is full of lies. The body tries to tell the truth. But,
it's usually too battered with rules to be heard, and
bound with pretenses so it can hardly move. We cripple
ourselves with lies."*

To summarize, then, there is a whole bouquet of
effects during a high which can all be profoundly useful
for the enhancement of your love life in various aspects.
During sex, you can experience the richness and uniqueness

of various altered states of consciousness which culminate in orgasm. All of these mind enhancements described in this chapter – and, arguably, many more – can thoroughly enrich sexual encounters and give a profoundly new meaning to the experience.[97]

The Endocannabinoid System, The Entourage Effect, and Cancellation Effects

There are numerous other fascinating questions in connection with cannabis and sex which will have to be better explored in the future. Are there some more direct physiological ways in which cannabis can work as an aphrodisiac? The endocannabinoid system seems to be involved in many sexual processes, so there are good reasons to believe that phytocannabinoids can have a deep and direct physiological impact on various sexual functions. Connectedly, we have reasons to believe that there is a huge therapeutic potential for phytocannabinoids concerning sexual dysfunctions if we look at the role of the endocannabinoid system.[98]

What, exactly, can a cannabis oil do for a woman if applied to her vagina? In which ways do we see different effects of the cannabis high on women and men, in general, and especially what concerns the sexual experience? Is it better to ingest cannabis products or to inhale them to get a better "body high" for the sexual experience? Which cannabis varieties are better for sex? Is it true that some cannabis varieties lead to vaginal dryness, while other varieties do not? How do cannabis lubricants work, and how can cannabis topicals affect your genitals? Can cannabis help to take away vaginal pain from various conditions and, thereby, help women to enjoy sexual experiences again, for instance after a trauma?

Despite a lot of anecdotal evidence of the aphrodisiac effects of cannabis and a lot of marketing claims, we are only at the beginning when it comes to seriously evaluating those claims and sorting out marketing and personal myths from evidence-based claims. At this point, though, it seems that it makes a lot of sense for users to experiment with various varieties and doses when using cannabis to enhance their love lives.

Users will have to find which cannabis varieties work best for them. Meanwhile, some cannabis breeders are striving to create the best 'sex-pot' varieties. We will have to consider their marketing claims with skepticism, of course, but it seems to make a lot of sense for consumers to begin an informed journey which also includes experimenting with varieties.

Again, this is my mantra: skills, experience, knowledge, and etiquette certainly play a decisive role. If you do not respect your partner's worries about your cannabis use or, for instance, if you force him/her to use it, the best variety of cannabis for sex will not help you anymore. If your dosing and communication go wrong, you may just become too anxious, too tired, or too introspective for sex. We live in a modern world in which we have almost forgotten about the knowledge of ancient cultures and their rituals embedding the use of cannabis. Yet, there are highly evolved traditions of using cannabis for sex that go back thousands of years.

Cannabis, Shiva, Yoga, and Tantra

In the region of the northern Himalayas and India, cannabis has been used since prehistoric times in shamanism. Reports about cannabis use in these regions

(17) A cannabis strain from Nepal's Mustang Valley in the experimental breeding stage.

go back as far as 4000 years ago, when the Indo Persians, also called *Aryans*, wrote their *Vedic* scriptures and started to bring their cannabis culture to India from the Northern Himalayan regions. In his landmark article "Tantric Cannabis Use in India" cannabis historian Michael Aldrich summarizes:

> *"The oldest religion on earth for which we have the complete texts was based on ritual ingestion of psychotropic drugs. The religion is that of the Vedas, texts composed in Northern India in the second millennium B.C[MU1] . (...)."*[99]

The oldest of the *Vedic* scriptures and one of the most important text sources for Hinduism, the *Rig Veda* (from

Sanskrit, Veda, 'knowledge', and '*Rig*', "verses') consists of 10 books. The total number of hymns is 10,552, out of which 1028 verses are about soma – the famous mysterious strong psychoactive drink they used for their religious ceremonies.[100]

Many plants and mixtures thereof have been proposed by researchers in the last decades as candidates for *soma* drink preparations, including *ephedra, Amanita muscaria* (*fly agaric*), *psilocybin* mushrooms, cannabis, and poppy seeds. Whatever *soma* was, it was certainly meant to be strongly psychoactive. *Vedic* priests famously sang:

"We have drunk Soma, we have become immortal, and we have attained the light, we have known the Gods."
[101]

Aldrich rightfully points out that the *"fountainhead from which sprang the Indo-European religions probably had as an essential element the ritual use of drugs."* [102]

The god *Shiva* is an amalgamation of older *Vedic* and Non-*Vedic* Deities[103]and inherits his phallic and adulterous character from one of his *Vedic* antecedents *Indra*, the Phallic God of Fertility.[104] According to one legend, *Shiva* wanders off after a dispute with his wife Parvati and awakes under a cannabis plant, tries it, and becomes rejuvenated after his exhausting journey. He is later described as choosing cannabis as his favorite plant and became known as *Bhangeri Baba,* the Lord of Bhang. The fourth *Vedic* scripture, the *Atharva Veda*, names the psychoactive cannabis drink bhang as one of the five sacred plants and calls it a source of joy and happiness, a liberator, and a substance that can help us to lose fear.[105] *Shiva, also,* is often symbolized in the form of a penis.

The ethnopharmacologist and shamanism expert

Christian Rätsch relates the *Vedic* story of how Parvati uses a cannabis drink to open *Shiva*'s third eye so that he can return to her and see her beauty and uniqueness again after all his adultery.[106] *Shiva* then rediscovers his sexual lust for his wife again and understands that she is the one for him. We have already mentioned awe and the freshness of experience as typical experiences of cannabis users during a high. This effect can play a powerful role in the love life, especially in a long-term relationship, and this seems to have been well-known thousands of years ago.

Robert Anton Wilson gave us an important evolutionary perspective on how fundamental this effect can be:

"(...) the human mother could not cope with the hazards and creatures of the wild (...) unless she persuaded a male to hang out with her. (...) Many forces can and have held families together, but the one that actually performs the lion's share of the job throughout human evolution is sex. (...) the human being is the only one without a limited mating season (...) Somehow, somewhere in evolution, the proto-human female mutated and leaped out of the estrus cycle of other apes. She was, so to speak, in heat all year long. And this persuaded the human male to remain with her all year long (...) and formed the foundation of human society – the primordial family".
[107]

The first Tantric scriptures stem from the 7th century A.D. but the *Tantric* tradition probably started earlier as a cultural amalgam of *Shaivite Hinduism* and *Tibetan Buddhism*.[108]

"Sex yoga developed concurrently with drug yoga in the late Vedic period. (...) The Kāma Sūtra and

*Ananda Ranga eloquently detail Hindu sexual
techniques, and the Tantras transform such sexual
practices into a means of meditational yoga.
Marijuana fits into sex yoga as well, for in Hindu
folk medicine it is the aphrodisiac par excellence ".
Shiva is the Lord of both asceticism and eroticism,
a seeming paradox about which volumes have been
written. (...) It is resolved in Tantric tradition by
the unity in practice (sādhana) of yoga with bhoga
(sexual pleasures). (...) Thus bhoksha (sexual
enjoyment) sublimated by rigidly prescribed Tantric
practice, produces moksha (release or liberation).
(...) Tantric practice brings together these elements
– the ceremonial use of marijuana, the conscious
employment of "poisonous" or dangerous substances,
and the practices of drug and sex yoga – into a fully
developed system for achieving mahānirvāṇa "*[109]

In Buddhism, *mahānirvāṇa is* the greatest nirvana,
the highest form of liberation, a transcendental state in
which there is neither suffering, desire, nor a sense of
self. Aldrich goes on to explain how important it was
for students of Tantric practices to go through rigorous
education to prepare for the highly sophisticated *Tantric* sex
ceremonies, and rituals including meditations, ritualistic
bathing, chanting, reciting mantras, and breathing exercises
during the onset of the cannabis high. One function of
these rituals certainly is to prolong the sexual experience
by delaying sexual intercourse and orgasm.[110] Note, by the
way, that the cannabis high can be used here in a seemingly
paradoxical way: it can intensify aphrodisiac effects and
a strong sense of being in the here-and-now which could
lead to premature ejaculation. Yet, it can also be used here
to hyperfocus on rituals that delay this process. In this way,
the high, intelligently employed, can play an important role

in building up tremendous sexual energy. Sexual pleasure is not only prolonged in real-time but, as we have observed before, an altered sense of time lets the Tantric practitioner experience everything in a subjective slow-motion.

The sãdaka (Tantra student) is asked to perform the ritual of Bhūta-shuddhī (the cleansing of the elements of which the body is composed) and to imagine *"an angry black man in the left side of the cavity of his abdomen the size of his thumb, with red beard and eyes, holding a sword and shield, with his hands held low, the very image of all sins "*.[111]

Again, we can see how the intensification and enhancement of our ability to imagine during a cannabis high come into play.[112] Obviously, thousands of years ago, it was already well known to Tantric practitioners how to use the cannabis high to enhance states of meditation and one's imagination to prepare for a divine and liberating sexual experience. Interestingly, also, Aldrich notes the intensity of the imaginary identification of the Tantric student with the divinity meditated upon:

"In the state of exaltation brought on by the marijuana and the concentration rituals, the Tantric "hero" does not regard himself as separate from the object of worship of the divinity meditated upon, he becomes it. " [113]

In the words of simulation theorists of empathic understanding, we could rephrase this as an intense imaginative melting into the simulated subject.

Aldrich concludes his article with a beautiful description of the role of cannabis in the Tantric sex ceremony: *"(...) in the ecstatic union of the human and the divine represented by this ritual, the sense of self is*

transcended by both partners. The role of cannabis in
Tantric ceremony is thus to enable the worshippers to feel
the divinity within and without themselves. " [114]

This seems to be quite the opposite state of that of my solipsistic acquaintance, who felt that without being under the influence of a cannabis high he would feel absolutely alone in this world.

Thousands of years ago, then, humans used cannabis for highly developed elaborate sexual practices to achieve altered states of consciousness, hyperfocused meditation, an extreme empathic connectedness, ego transcendence, and a state of unique bliss through ecstatic sexual experiences way beyond what is conceivable to most of us today, as the ultimate liberating experience. They knew how to control dosage, set, and setting, and they invented rituals perfectly designed to unfold the potential of cannabis in the sexual experience. We can safely assume that they were well aware of many of the mind-enhancing effects of cannabis that I have described in this book.

I sometimes wonder if thousands of years ago, some tantric herbalists also tried to create the ultimate cannabis variety and concoctions thereof to best come to enlightenment through a sexual experience.

As I am writing this chapter, there is already a thriving market for cannabis products for the enhancement of sex in the U.S., Canada, and some other countries. We will probably soon see millions of more cannabis users with an interest in the cannabis high for the enhancement of lovemaking. And as the market will grow, tens of thousands of yoga teachers, sex therapists, and many other interested professionals will integrate cannabis into their practice.

There is a blissful and ecstatic journey ahead for all of us – a journey that began thousands of years ago.

(18) Another experimental variety based on the well-known Jack Herer variety.

VIII. Mind Enhancements and the Endocannabinoid System

"The endogenous cannabinoid system – named for the plant that led to its discovery – is one of the most important physiologic systems involved in establishing and maintaining human health. Endocannabinoids and their receptors are found throughout the body: in the brain, organs, connective tissues, glands, and immune cells. With its complex actions in our immune system, nervous system, and virtually all of the body's organs, the endocannabinoids are literally a bridge between body and mind."

Bradley E. Alger, *Getting High on the Endocannabinoid System* (2013)[115]

Why does the plant cannabis affect so many of our bodily and mental functions? Why has it been used for millennia in so many cultures around the world medicinally for such a wider range of medical conditions? Some thirty years ago, researchers discovered cannabinoid receptors in humans that interact with the plant cannabinoid THC (delta-9 tetrahydrocannabinol). It was clear that the human body would not build receptors only for the vague chance of a human consuming cannabis, so the hunt was on for *endogenous* ligands, substances synthesized in the human body itself that would specifically bind to those receptors and, thereby, activate certain functions. Soon researchers would find this ligand and called it *anandamide* (Sanskrit for "bliss"). Further research uncovered that cannabinoids play a unique role in the human body as scientists began to discover a biochemical human endocannabinoid signaling system (ECS). The human body, they found, produces its cannabinoids, the endocannabinoids anandamide (AEA), and 2-arachidonoylglycerol (2-AG). These endocannabinoids bind to endocannabinoid receptors called CB-1 and CB-2. These endocannabinoids and their receptors along with enzymes synthesizing, transporting, and degrading them function in an endocannabinoid signaling system in our brain and body.[116] The endocannabinoid system also includes enzymes like the fatty acid amide hydrolase (FAAH), which breaks down anandamide. Later, they found more endocannabinoids, including 2-AG ether (2-AGE), and capsaicin-like N-arachidonoyl dopamine (NADA).

Since its initial discovery scientists have published more than ten thousand scientific articles on endocannabinoids. We now know that the ECS is responsible for the control of a whole variety of cognitive and physiological functions in humans. Most researchers in the field believe that the ECS is a *homeostatic* (balancing) system, maintaining health in our body and mind. All humans have it, and almost all animals, except for insects. Endocannabinoids and their receptors are built in the human body independently of any previous contact with phytocannabinoids from the plant cannabis. Despite its importance for the regulation of a wide variety of physiological and mental functions in humans, most medical professionals in the world have never heard of it, as it was not part of their education.

The Many Functions of the Endocannabinoid System

Evolutionary speaking, we now know that the ECS is approximately 600 million years old, as endocannabinoids are found also in sea squirts, organisms that have appeared on the evolutionary scene around this time and still exist in our oceans today. This means that the ECS appeared on the evolutionary scenery long before the *Cannabaceae* family and the cannabis plant arrived on the scene some 30 million years ago.[117] Surely, then, the success of the cannabis plant has to do with its ability to interact with the endocannabinoid system in animals.

The worldwide triumphal procession of the cannabis plant in the last centuries has to do with the fact that cannabis is the only plant that can express THC, CBD (cannabidiol), and more than 140 other cannabinoids as well as a whole range of terpenes and flavonoids. The cannabinoids of the plant interact with our cannabinoid receptors and can therefore influence the functions of this system. Furthermore, we now know endocannabinoids are "promiscuous" and do not only interact with CB-receptors but also with other endogenous receptor pathways, such as TRPV-receptors, nuclear receptors like PPAR-α and PPAR-γ, Orphan GPCRs (G-protein coupled receptors) as well as GABAA and other receptors. Likewise, phytocannabinoids can interact and influence other receptor pathways than the endogenous cannabinoid CB receptors CB-1 and CB-2.

The last decades of research on the endocannabinoid system have shown that the CB-receptor pathways are involved in the regulation of many functions, such as the regulation of mood, perception, cognition, fear, stress, memory, sleep, appetite and food intake, visual development and locomotion, neuroprotection, control of body temperature, and immune system responses. The endocannabinoid system is also involved in the activation of the immune system, pre-and post-natal development, embryo implementation, and fertility, among other functions.[118]

Other receptor pathways influenced by endocannabinoids as well as many phytocannabinoids are for instance involved in *nociception* (pain sensation), perception of itch, thermal perception, anti-inflammatory actions, chronic pain, atherosclerosis, energy balance, sedation, and epilepsy. The psychological effects of THC and some other cannabinoids can to a large extent be explained by its binding affinity to CB-1 receptors. These CB-1 receptors are widely distributed in the brain:

> *"Briefly, CB1 expression has been reported in various regions from the spinal cord to the neocortex, with prominent expression in nociceptive fibers, the retina, olfactory bulb, striatum, amygdala, hippocampus, prefrontal cortex, cerebellum, brainstem, and hypothalamus. CB1Rs are present on various cell types including select excitatory (glutamatergic) and inhibitory (GABAergic) neurons, astrocytes, and potentially microglia and oligodendrocytes. (...) (T)he possible functional role of presynaptic CB1 is in regulating neurotransmitter release, while postsynaptic CB1Rs may modulate neuronal excitability. CB1 is also found on intracellular membranes, most prominently mitochondria, where it may play a role in regulating energy homeostasis and ultimately synaptic transmission."*[119]

Even if you are not a neurobiologist, this passage alone should make it clear how ubiquitous the distribution of endocannabinoid receptors is in our brains, and in how many ways cognitive and other functions the endocannabinoid system must be involved.

The important homeostatic role of the endocannabinoid system and the functional connection to other physiological regulating systems explains why we have seen such a wide range of therapeutic uses of the cannabis plant for so many medical conditions throughout human history, for more than 4000 years – and probably much longer – in ancient cultures such as China, India, Persia, Egypt, and in many other countries around the world.

In this book, I have discussed a whole variety of mind enhancements cannabis users can experience under favorable conditions when consuming cannabis. It is obvious that the manifold effects of the plant on our mind and body are mediated to some degree by affecting an already existing endocannabinoid system which controls a whole variety of mental functions. The endocannabinoid system also seems to be involved in highly developed homeostatic mechanisms, for instance as it concerns how we deal with traumatic memories. Famously, it has been shown that the endocannabinoid system plays a major role for us when it comes to dealing with negative or traumatic experiences. The storage of aversive memories in animals and humans is one of the most important functions mediated in the central nervous system. The endocannabinoid system plays a role in helping us to gradually over time get over traumatic experiences.[120] This could explain why phytocannabinoids in medical cannabis seem to help veterans and others afflicted with post-traumatic stress syndrome so well.[121]

For decades, research on cannabis and cannabinoids was made almost impossible in the U.S. because of its classification as a Schedule I drug, allegedly a substance with no currently accepted medical use and a high potential for abuse. A lot of the funding for studies went into researching the risks of cannabinoids rather than their potential. I will not argue the point in detail here, but I would like to inspire researchers from the relatively new field of agnotology to take a very comprehensive look at the research on cannabis and cannabinoids in the last decades. *Agnotology*, from the Neoclassical Greek word agnōsis (ἄγνωσις, 'not knowing'; cf. Attic Greek ἄγνωτος, 'unknown') is a relatively new discipline within the sociology of knowledge that specifically studies the deliberate and cultural generation of doubt or ignorance. Stanford university professor for the history of science Robert N. Proctor and the social historian Iain Boal, who coined the term *"agnotology"* in 1995, argued for instance that the tobacco industry ran a disinformation campaign for over 40 years producing doubt about the already scientifically well-known facts about the detrimental health effects of tobacco smoking. In their landmark book *Merchants of Doubt: How a Handful of Scientists Obscured the Truth on Issues from Tobacco Smoke to Global Warming*, authors Naomi Oreskes and Erik M. Conway go on to argue that the campaign techniques of the tobacco industry were then used by the fossil fuel industry to argue against the overwhelming scientific evidence for a man-made climate change.[122] I think it is about time that agnotologists take a comprehensive look at the cannabis prohibition since the early days and its dynamics.

For now, let me just point out that it was and still is certainly rather difficult for scientists to look into the positive potential of phytocannabinoids as it concerns their effects on higher cognitive functions. Although the endocannabinoid system seems to be a highly interesting target for pharmacological therapeutic interventions with cannabinoids, the interest of most pharmaceutical companies to finance studies is limited because the plant and its phytocannabinoids cannot be easily patented. In a letter to the NFL in 2014, Lester Grinspoon asked the organization to get involved in the funding of research – research that would cost millions of dollars but would also be in the self-interest of the NFL because CBD is without question neuroprotective and could for instance help players with concussions.[123] I guess it shows something if one of the most eminent experts of medical cannabis in the world gives a shout to the NFL to finance studies.

Currently, it seems to be quite obvious to many researchers that the ECS is involved in supporting many cognitive functions in the human brain, given data from animal studies and clinical data on the influences of the ECS in neurodegenerative diseases. The latter might be a strong driving force for scientists to further investigate the ECS:

"Analysis of clinical data on the influences of ECS activation in neurodegenerative diseases, along with experimental data obtained in animal models of these diseases, in most cases points to a positive role for EC in brain operation and particularly its cognitive functions. (...) The protective role of EC is clearly apparent on development of neurodegeneration, especially at the early stages of pathological changes, indicating an important role for the EC system in activation of the protective mechanisms of the brain."
[124]

Presently, some researchers also believe in the hypothesis that several medical conditions like migraine, depression, multiple sclerosis, Parkinson's disease, fibromyalgia, and irritable bowel syndrome, could be caused by what has been called *"Clinical Endocannabinoid Deficiency"* (CED) syndrome.[125]

(19) Close-up of the approximately 0,5 inch long stigma of a Silver Haze cannabis variety.

The Effects of Phytocannabinoids on the ECS

Despite a growing body of evidence from patients, studies, and animal models, it would yet need to be researched in which ways phytocannabinoids would actually help for each of these deficiencies. So at this point, even if we believe that many neurodegenerative and other diseases could be explained by CED, we need to be cautious to make claims about the potential of phytocannabinoids in their treatment, as we still need randomized controlled studies to evaluate various therapeutic cannabinoid medication options.[126]

What about the acute effects of phytocannabinoids on our endocannabinoid system and the cognitive functions it regulates? Many clinical studies come to contradictory and mostly negative results when it comes to the results of phytocannabinoids acting on cognitive functions such as episodic memory or attention controlled by this system, which probably has to do with a certain degree of ignorance concerning influential factors such as dosage, methods of administration, differences in tolerance of the subjects, as well as set and setting.[127]

As we have seen, though, a thorough analysis of hundreds of reports of cannabis users suggests that many of the highest cognitive functions we know may be affected by cannabis use; some of them positively, under certain conditions, but we also have many reports about negative effects. Users have reported effects on their attention (redirection of attention, hyperfocus of attention, difficulties with sustained attention), working memory (usually negative reports about "losing the thread" during a conversation), episodic memory (more detailed, retrieval of memories that usually seem inaccessible, etc.), a greater acuity in perception, synesthetic experiences, intensified imagination, mind racing (a speeding up of associative streams of thought), various kinds of distortion of time perception, and enhanced body perception, enhanced pattern recognition, mood modulation, enhanced introspection and empathic understanding, altered and often (but not always) enhanced sexual experiences, a severe positive or negative impact on creative abilities, and on our ability to arrive at deep, spontaneous insights.

Hypothesis: The ECS is multi functionally involved in all hierarchy levels of our higher-level cognitive abilities (IN-ALL-HI-hypothesis)

Now, again, as in the case of the endocannabinoid deficiency hypothesis, we have to look at two questions distinctly:

(a) Is it possible that the endocannabinoid system is deeply involved in many of these cognitive functions, like attentional focus control, or pattern recognition?

(b) Can we temporarily positively affect functions of the endocannabinoid system with phytocannabinoids under certain conditions?

Let me give a brief answer to (b). As we have seen, many reports of adult users from the inspirational realm, as well as from patients using cannabis, for instance for ADHD, suggest a positive impact of exocannabinoids (like phytocannabinoids or other synthetic cannabinoids brought into the system from outside) is indeed possible under certain conditions. Users have to be able to control their individual adjusted dose in certain settings, adjusted to their mindset and other factors. In other words, users need to learn how to "ride a high". Given that many consumers of cannabinoids do not know how to dose cannabis or use it correctly, we can explain why so many users have experienced various negative effects from cannabis on their cognitive functioning during a high.

What about question (a)? Could it be that the endocannabinoid system is functionally involved in the control of the highest cognitive functions known to us, including pattern recognition, creativity, introspection, empathic understanding, and insight? If yes, could the endocannabinoid system be a road for basic science to investigate the multilevel neuronal functional architecture of these mental processes and abilities?

To come closer to an answer to this question we first need to understand that our current understanding of the neurological processes underpinning those higher cognitive functions is rather limited. We have seen the limitations of neuroscientific theories concerning this level of higher cognition in our critical discussion of the mirror neuron system as the suspected underlying system of empathic understanding. If we want to understand better how the endocannabinoid system could be involved in higher cognitive functions, we need to first understand that we are going into this new terrain. This is exactly why I believe that researching the endocannabinoid system under the aspect of defining those higher cognitive states is such an interesting enterprise – and why it could be a decision that entirely changes the game of understanding the human mind.

For understandable reasons, endocannabinoid scientists in the next decades will be driven mainly by medical interests and needs – that's where the funding will take them. I am hoping, though, that my research gives some basic researchers from the cognitive sciences and philosophy more initiative to investigate the human mind in the light of the yet to be explored role of the endocannabinoid system in controlling many higher cognitive processes. I am confident that this basic research will lead to results from which patients and many others will profit; it will lead to a deeper understanding of human thinking and behavior, and this will help scientists in many areas to come to a better understanding of the healthy and the sick human mind.

For now, let me distinguish two ways in which the endocannabinoid system could be involved in these most complex processes of higher cognition:

(a) The Endocannabinoid System is involved in *regulating processes of lower-level cognitive states*, which then affect higher-level states in which they are fundamentally involved

(b) The Endocannabinoid System is involved in the *cognitive architecture of higher-level abilities itself*, helping to orchestrate the lower-level states

Let me explain these suggestions. Cognitive scientists usually see cognitive processes in a hierarchy. More complex cognitive abilities like empathic understanding depend on and involve many perceptual and cognitive abilities which can be seen on a "lower-level", such as pattern recognition, episodic memory retrieval, and imagination. Your pattern recognition of a "sadness" in your friend's voice will help you to better understand him, or your visual recognition of his "happy" walking style. It will be easier for you to empathically understand your 10-year old kid's joy and pride about finally managing to throw a boomerang and catching it if you can retrieve episodic memories of your childhood. If you want to understand the panic of your neighbor who just almost fell off his roof and is now hanging with one hand from the edge, your ability to imagine how this would feel helps you to empathically understand him. Briefly, then, empathic understanding is one of our most developed skills, period, and it involves many other lower-level cognitive abilities, such as pattern recognition, episodic memory retrieval, and imagination. Enhance one of these abilities and you can see how your empathic understanding may become better. This would be hypothesis (a): The Endocannabinoid System regulates pattern recognition, episodic memory retrieval, and imagination, and therefore, changes in those processes would also lead to changes "bottom-up" to a higher-level ability to empathically understand others. In my view, these cognitive bottom-up processes obviously play a role when we look at cannabis enhancing higher cognition. Also, we have good reasons to believe that the endocannabinoid system is deeply involved in the regulation of processes such as pattern recognition or imagination – which would explain why we are seeing effects on higher cognitive abilities, like empathic understanding.

Now, to hypothesis (b): Could it be that the endocannabinoid system is involved in the architecture of higher-level processes itself, over and above affecting lower-level cognitive abilities? Higher cognitive abilities like empathic understanding need not only affect lower-level processes like episodic memory to be involved – these processes need to be balanced and "orchestrated", and this controlling function needs to be manifested in the higher-order neuronal network architecture of our brains

As we have seen, it has already been shown that the ECS is primarily involved in homeostatic balancing functions, and also in so many functions of the human mind. Complex cognitive processes like attentional focus need various highly intelligent balancing functions. A strong attentional focus can help us to solve a difficult task, but if we are not diverted by a tiger in the fringe of our perceptual field, he may attack and eat us. To survive, we need to keep our attention balanced; we should be able to focus, but we should still get distracted by certain stimuli in our visual field.[128]

Could it be, then, that the endocannabinoid system is involved in balancing the more complex architecture of neural networks which seem to enable us to integrate and control several other lower-level cognitive abilities to, for instance, empathically understand others? Scientists interested in the architecture of the most complex cognitive abilities of humans should find new, interesting roles of the endocannabinoid signaling system in this architecture. Of course, these functionalities of the endocannabinoid system need not be restricted to "balancing" tasks. Could it be that it is functionally involved in other ways in processes of higher cognition, such as the orchestrating of various cognitive functions for introspection, creative thinking, or empathic understanding?

My conclusion would be that it makes sense to hypothesize at this point that the endocannabinoid system is involved in a multitude of regulatory functions on all levels of cognition, even at the highest level we know. The multidimensionality of the cannabis high, then, would be a hint at the multi-functional involvement of the endocannabinoid system in all or several hierarchy levels of our highest cognitive abilities.

These questions take us deep into the realm of basic science, offering a possible road to a deeper understanding of the human mind itself. The enterprise involves an interdisciplinary approach of philosophers of mind, cognitive psychologists, neuropsychologists, cognitive linguists, evolutionary biologists, pharmacologists, and several other groups of scientists. It would be a huge interdisciplinary effort of many scientists to research the effects of cannabinoids and to further develop hypotheses for functions of the endocannabinoid system in higher-order cognitive functions such as episodic memory, control of attention, bodily sensation, pattern recognition, imagination, or even the control of introspection, empathic understanding, creativity, or our ability to come up with deep and spontaneous insights. Let me be clear: I have given up believing that it is realistic to find government funding for such research. When it comes to initiating an interdisciplinary approach to researching the endocannabinoid system's role in human higher cognition, our best bet is probably a cannabis aficionado billionaire who is convinced that researching the ECS for these matters can add something to their brand image.

Let us take a perspective from a possible future in which we can see that the endocannabinoid system indeed plays a major role in the highest cognitive abilities of humans, including creativity, introspection, empathic understanding, and creativity. Would it not be ironic to see that for more than a hundred years, humans have outlawed the only plant that can stimulate exactly the abilities we consider to be the core and defining human mental abilities?

Balance, Imbalance, and Mind Enhancements

If the endocannabinoid system is a balancing system for various physiological and mental functions to the extent that has already been established by scientists, then it makes intuitive sense that consuming cannabis could help those with an imbalance caused by an endocannabinoid deficiency, for instance. Yet, assume I am healthy and have a normally functioning endocannabinoid system. If I consume cannabis without any deficits in my endocannabinoid system and flood my system temporarily with cannabinoids by inhaling from a vaporizer to get "high", would that not mean that I am causing a state of imbalance? How can this be healthy, or enhance my mind?

I guess the short answer is that we all need balancing functions, but we also need to "oscillate" between various states of consciousness. Altering our state of consciousness is deep in our nature. Every day, circadian physiological mechanisms in our brain force us to fall asleep and dream – wild, altered states of consciousness, like a trip on a psychedelic substance. We go through states of ecstasy when we have sex or celebrate victories in crowds, we fall into trance states when we dance or drum, we experience fight or flight or freeze mode in situations of stress and danger, where we hyperfocus our attention on an expected danger. Many of us tend to make the mistake of rationally reconstructing our human existence as a state of rational wakefulness. This is not who we are. Altered states are part of our nature, and we use them productively to survive in this world. Our fight-or-flight mode helps us to focus extremely, and the adrenaline kicking in helps us to stay awake, to be alert, to forget about the pain of fighting wounds, or of our muscles aching from running from a predator. This mode can be very counterproductive in many modern situations which may trigger it, like giving a presentation at school. The cannabis high, also, is an altered state of mind which is characterized by a marked difference from the "balanced" state of wakefulness. Our attention, for instance, is often hyperfocused, which can be unproductive and even dangerous in some situations. However, if we learn how to include this altered state of consciousness productively into our lives, then it can be incredibly resourceful, just as it can be useful for us to remember and integrate our dreams into our lives.[129]

(20) Close-up of the upper part of a Silver Haze flower. The mushroom shaped trichomes of the plant are still clear and not ready to be harvested yet. It makes a big difference for the cannabis high when in the maturation stage the plant is harvested and how the plant material is treated in the curing and fermentation process, as its chemistry changes during these phases. Many illegal growers tend to let their plants mature too much before harvest because the plant gains more weight, and thus, more profit.

IX. Cannabis, Coping, and the Stigma of Addiction

"We must learn to regard people less in the light of what they do or omit to do, and more in the light of what they suffer."
Dietrich Bonhoeffer (1906-1945),
Letters and Papers from Prison

"As long as habit and routine dictate the pattern of living, new dimensions of the soul will not emerge."
Henry van Dyke Jr. (1852 –1933)

Addicts. They make me nervous sometimes. And they seem to be everywhere. Frantically, they move around in circles, some of them slave to implemented passions, others manipulated, in self-denial, and helpless.

At six in the morning, the first class of addicts runs to get their daily fixes: the classic workaholics. They swarm out of their houses to go and work like maniacs. Some are craving for more power, suppressing the persistent feeling of having slowly turned into corporate marionettes. Others work excessively to satisfy their shopping addiction and buy whatever they have been told makes them better, happier, or more valuable people. Most of them are multiple *drug addicts* with a long history of abuse: hundreds of millions are addicted to several caffeine shots in the morning and afternoon, or they need their drug sugar, and its instant hit almost hourly. Others are addicted to nicotine and the dozens of other psychoactive chemicals in cigarettes which give them a short-lived dopamine high and staves off their inherent depression for just a few minutes before the craving starts all over again. A growing number of people are physically addicted to alcohol or other depressants, sedatives, opioids, other painkillers, benzodiazepines, stimulants like amphetamines, or alkaloids like cocaine, or are psychologically dependent on selective serotonin, noradrenaline, or dopamine reuptake inhibitors, or various other new antidepressants. And that is only the beginning of a much longer list of substance addictions.

A more recent class of addicts are what I like to call the *cybersocioholics* – those unhealthily obsessed with abundant yet meaningless communication through a multitude of social networking tools and other communication channels. Compulsively, they check their three or four email accounts, then go to their Facebook mailboxes and pin walls, chat a little while constantly looking up and responding to their WhatsApp, Telegram, or Signal messages on their smartphones, then maybe an occasional glance at their LinkedIn accounts, then switch and check what happened on their blogs or in their internet dating portals. Once they finish dealing with the last of their 10-15 communication channels – not counting landline telephones, one or two mobile phones, or their post boxes – they return to the start of the loop. It's been almost an hour . . . maybe somebody finally posted something new on Facebook? Maybe somebody replied to my new post on Instagram?

Naturally, the myriads of cybersocioholics are only a subgroup of the larger group of the *internetaholics*, just like the *cyberpornaholics, cyberpokeraholics,* and other *cybergameaholics*. If their internet activities would not be so compulsively restricted to satisfying their respective addictions, they might stumble upon the website *netaddiction.com*, which offers treatment services in the categories "Cyberporn/Cybersex, Online Affairs, Online Gambling, Online Gaming, Compulsive Surfing, eBay Addiction."

Other media addicts are sticking to a more old-school addiction to television, Netflix & YouTube video streaming, or computer games. On TV or in the movie theater they watch action movies or war movies, car races, or professional wrestling to kick up their adrenaline levels like nothing else in their safe and streamlined *real* lives does anymore – and so one way or another they spend half of the time in simulated virtual realities. They belong to the larger class of what we could call *virtuaholics*, which includes everyone addicted to the ever-growing virtual world, a safe and controllable refuge helping them to escape from reality.

Long before the rise of the internet and mobile phones, Susan Sontag described in her 1977 essay collection "On Photography" how we all have become addicted to images:

> *"Needing to have reality confirmed and experience enhanced by photographs is an aesthetic consumerism to which everyone is now addicted. Industrial societies turn their citizens into image junkies; it is the most irresistible form of mental pollution. (...) It would not be wrong to speak of people having a compulsion to photograph: to turn experience itself into a way of seeing. Ultimately, having an experience becomes identical with taking a photograph of it, and participating in a public event comes more and more to be equivalent to looking at it in photographed form."*[130]

Wiiholics are the blessed ones in the group of *virtuaholics*: to outsiders, they may look ridiculous jerking their limbs around in a weird way, but the industry made a nice little concession to their human nature by encouraging them to move around a bit while perceptually hooked up to their virtual reality, whereas most other *virtuaholics* are usually sitting there like frozen dummies, perceptually hooked to screens showing strangely impoverished glowing colors. Many of them seem to have lost touch with their bodies after suppressing the feeling of self-inflicted paralysis for years. Yet, their bodies come out of a long evolutionary process that predestinated them to run, hunt, and fight, so now they are all starting to feel uncomfortable and becoming addicted to Ritalin for their attention deficit hyperactivity disorder. Their doctors are usually part of a "health care" system generating money by turning people into prescription drug addicts rather than improving their health. And so they are handed prescriptions for all kinds of tranquilizers and sleep-inducing drugs to force their bodies to keep quiet even after a virtual day in subconscious paralysis on a stylish wheelchair. This section of the health care system is only a part of an entire industry designed to nourish and satisfy the addictions of its consumers – reinforced addictions are magnificent moneymaking machines. And make no mistake, newer virtual reality applications will change the game entirely – for many, the newly constructed worlds with ever-increasing levels of details, mystery and fun will be hard to resist.

If I look at the list of addictions named above I feel inclined to say: guilty, my Lord! I am struggling with several of the addictions described above! Then, again, to be honest, I do not feel guilty at all. I feel like a typical human. I belong to the vast majority of humans who are struggling from time to time with one or several of the addictive behaviors described above.

General signs of addiction are usually minimally described as follows:

"Addiction is an inability to stop using a substance or engaging in a behavior even though it is causing psychological and physical harm."[131]

Think about the addictions described above and ask yourself if you have been free of any of those addictions in the past; and which of your addictive behaviors you are fighting with most currently.

If we all understand how vulnerable we are when it comes to addictions we may become more cautious about pointing our fingers at those struggling with substance abuse disorders (SUDs). The third president of Germany, Gustav Heinemann, once famously said: *"Those who point their index finger at other people should never forget that three fingers of their hand are pointing at themselves."*

Maybe what makes me and many of us most nervous about watching addicts is that deep down, we know that we are all struggling with addictive behaviors – at least to some degree. We need habits to run our lives and many of these habits are "addictions" upon closer examination; it is hard for us to stop them even if we perceive that they are causing harm in some form.

Getting Into the Habit of Breaking The Habit

Sometimes I feel like I need a break from all these addictions and obsessions around me. I then go to Portugal to participate as a player at the famous Bar do Peixe Ultimate Disk beach tournament, chase a plastic disk all day long, breathe fresh air, and jump in cool Atlantic water. In the evening, I sit down on top of a dune at the magnificent Meco Beach one hour by car south of Lisbon, use a vaporizer to inhale some fine variety of cannabis, and watch the light slowly fade away after the sunset.

The cannabis high takes me in the here-and-now, and my mind frees itself from the daily hustle, needs, and plans. I breathe. Time is slowing down. I focus on the immediacy of my experience and feel my deliciously tired body, becoming more aware of my muscles, aching from running all day long. I feel alive, the warm wind on my face, and smell the salty scent of the sea. In this expanded headspace, I begin to feel free as if standing on a tower, looking down on my life. My high enables me to perceive patterns in my life and the lives of others, including our routines, addictions, and obsessions. Vivid memories come up and I can relive episodes long past. The high helps to transport me back into my past. With my enhanced episodic memory, I mind-race through previous experiences, I can see personal developments, stand outside my routines and addictions and identify them as such. I realize that many of the addictions described above, like cybersocioholism or virtuaholism, afflict me, too. But now, my mind is flying, and my attention doesn't seem to be drawn away by the usual forces. I feel free like a bird and momentarily liberated from my addictions. My enhanced introspective abilities enable me to identify and describe my addictions, to become aware of them as unhealthy patterns in my life, habits that keep me on a trajectory I do not want to follow. The cannabis high helps to break the routines, interferes with my addictions, helps me to step outside of the box, to make conscious decisions on my future behavior, and to get sudden insights. Associations come quickly. New notions occur to me, like "cybersocioholism", based on my enhanced ability for pattern recognition during a high.

Do not get me wrong: surely, the fact that cannabis can be used as an antidote for other addictions does not mean that it cannot itself be addictive. Cannabis may be generally much less addictive and generally, much less toxic than other substances, even legal substances, such as alcohol or tobacco, and the addiction may be primarily psychological. But even a purely psychological addiction can be destructive and unhealthy. So, well, surely there is a risk of addiction, too. Nonetheless, there is also a risk of getting unhealthily addicted to sex, car driving, or social networking – that does not mean that all these activities inherently bring us harm or have to be prohibited. All these activities can bring us great advantages, joy, and meaning in life if we know how to make good use of them.

(21) A tiny dewdrop sits between the stigmas of a Silver Haze flower growing from the so-called calyx (lower right).

Addiction and Coping Behaviors

If we want to come to a better understanding of how to avoid or deal with a cannabis addiction, we first have to better understand what an addiction really is and how it can be caused. It seems that many people are still clinging to an almost ridiculously primitive view of addiction, especially when it comes to substance-abuse disorders (SUDs), thinking that there is some kind of "evil" property of a substance that alone is responsible for turning healthy people into addicts.

Usually, however, healthy people do not become addicted because they try a substance out of curiosity. Most people who struggle with substance addiction, or, better, a substance abuse disorder, had pre-existing problems with pain, trauma, or other mental or physical problems for which they were seeking some sort of relief. As the addiction expert Gabor Maté writes:

> "Not all addictions are rooted in abuse or trauma, but I do believe they can all be traced to painful experience. A hurt is at the centre of all addictive behaviors. It is present in the gambler, the Internet addict, the compulsive shopper, and the workaholic. The wound may not be as deep and the ache not as excruciating, and it may even be entirely hidden – but it's there. As we'll see, the effects of early stress or adverse experiences directly shape both the psychology and the neurobiology of addiction in the brain."[132]

Importantly, Maté adds to this:

> "It is impossible to understand addiction without asking what relief the addict finds, or hopes to find, in the drug or the addictive behaviour."[133]

So, what kind of relief are cannabis addicts looking for?

I think we should re-phrase this question: How do these addicts try *to cope* with their traumas or other problems when they are using cannabis? Let me explain why I think this rephrasing matters.

Some years ago I attended a conference on the relationship between mobbing (bullying of an individual by a group), self-harm, and suicide among adolescents. An expert on mobbing behaviors[134] explained an interesting reaction of many adolescents to mobbing experiences. He told his audience that in phases of depression or distress from this trauma, adolescents would use knives and other sharp objects to cut and hurt themselves. He warned psychotherapists to simply forbid their young patients to engage in such self-harming behaviors. He suggested a therapist should at first tell those adolescents that they have effectively found a way to cope with their depression and auto-aggression in a way that probably helped them to survive. In an episode of extreme emotional distress, self-harm may distract an adolescent from the emotional pain, and it could help to "wake up" from a state of numbness. It has been shown that during the pain of self-harm, endogenous opioids are released in the body, diminishing the pain and leading to pleasure – both mental reactions that may help to resist the inner urge to commit suicide in an episode of uncontrollable emotional turbulence.[135]

Now, of course, self-harming behaviors as a coping strategy are not what an adolescent should opt for in the long run. Often, some of these self-harming adolescents seem to be even "addicted" to this behavior. They have formed the habit of self-harm because they have learned that it can help them in those distressful moments. This behavior, however, obviously leads to more problems. Apart from the injuries, which may be quite severe, self-harm could add to the feeling of shame of those adolescents, shame that they may already feel for the initial trauma. This leads to further social isolation and more problems. A therapist, therefore, has to approach this self-harming behavior carefully, to slowly help to give his patient better and more healthy coping strategies as she is trying to deal with the underlying trauma or pain, or some other underlying reason for this coping behavior.

What can we learn from this coping story about other addictive behaviors, or, in particular, what can we learn from this about "cannabis abuse disorder"?

Throughout human history, humans have used cannabis for a huge range of therapeutic applications, such as for conditions of severe and chronic pain, stress, anxiety disorders, nausea, epilepsy and seizures, sleeping disorders, Crohn's disease, ADHD, PTSD, and many more. We are beginning to understand better how cannabis and cannabinoids act on the endocannabinoid system to mediate a whole spectrum of therapeutic effects. As many countries have changed their legislation towards cannabis and cannabinoids in the last years, millions of patients have gained access to medical cannabis treatment, and even more have gained access to cannabis for responsible adult use.

It seems obvious now that millions of cannabis users who lacked and still lack legal access could not find the right relief for their medical conditions in other medications. As we accumulate evidence about the effectiveness of medical cannabis and cannabinoids, we can now say with confidence that many of them have found a very effective way to self-medicate with cannabis, for instance when looking for relief from chronic pain.[136] They have found a coping strategy for a whole range of problems – not only medical problems but, arguably, also some existential problems.

For many, cannabis use has become a coping strategy for surviving the stress of an out-of-control neo-liberal and de-solidarized society driven by the greed of oligarchs who demand their consumers and workers to "function", which is, to obsessively produce and consume. And those who "function" best for that end are addicted to work and consumption, and the images of movies and advertising.

Cannabis can indeed be a very effective temporary help to deal with this kind of stress. It helps you to get back to feeling your body, brings you to the here-and-now, helps you forget about the stress for a while, to feel awe, gain introspective insights, and become aware of unhealthy habits. And there are many other ways in which cannabis may help to relieve the stress you are feeling.

So, is cannabis a good coping strategy and self-medication strategy for all of us and a whole spectrum of problems and medical conditions? Or should it be replaced by better strategies to deal with medical or other existential problems? The answer to this question is, of course, that we need to look at individual cases. For some users, cannabis may be a very good option, even in the long term, whereas for others, cannabis use could be a coping strategy that needs to be replaced or at least minimized in its role. For those who need help with post-traumatic stress, for instance, cannabis may help substantially, over a long time. As mentioned above, we have good evidence from studies that the human endocannabinoid system plays a crucial role in diminishing the emotional stress resulting from traumatic experiences. We also know that many of those afflicted with PTSD have self-medicated with cannabis, using exogenous cannabinoids to modulate their endocannabinoid system to better deal with post-traumatic stress.[137]

Importantly, though, cannabis users will have to make an individual decision which will depend on an evaluation of other, alternative treatments like psychotherapy, psychoanalysis, hypnosis, other pharmacological interventions including psychedelic options, or activities like meditation. Also, they will have to find out about cannabis varieties, dosing strategies, and which methods of consumption that work best for them.

In the current situation, we must assume that many patients pay a high price because they neither have good access to good sources of information about cannabis nor do they have access to well-controlled cannabis products on a legalized market. They do not understand which varieties work best for them, which methods of consumption are best for them, and they often do not know how to approach dosing. Many only have access to cannabis of poor quality from an illegal market, which makes them too dazed and confused to mentally function the way they like to function daily.

Sadly, the biggest price most people pay for using cannabis to cope with problems like PTSD has to do with the stigma around cannabis and the prohibition that has been reinforced for such a long time. Given our knowledge concerning medical cannabis today we can say that for many cannabis users worldwide, what is often classified as a cannabis addiction could well be a good coping strategy and for many an excellent medical treatment option. As long as there is a prohibition and as long as the stigmatization around cannabis use remains, this coping strategy leads to completely unnecessary and grave negative consequences. Cannabis users have to fear being criminalized, stigmatized, socially marginalized, and discriminated against because of their use of an illicit substance.

Stigmatization or even criminalization of those who are using or abusing psychoactive substances like cannabis or opioids for some kind of suffering usually leads to social isolation, economic failure, and more shame – which means more human suffering, and it all spirals down from there.

Cannabis as a Cure for Substance Abuse Disorders

Unquestionably, some substances like the class of opioids are more addictive than others in general and more addictive than cannabis, specifically, in the sense that they have a different impact on our brain chemistry, especially if they are directly triggering our neurological reward systems. Opioids bring a completely different set of problems than cannabinoids. They cause us to build a stronger tolerance, lead to much more severe withdrawal symptoms, have severe side effects, are problematic in what concerns drug interactions, and can be deadly if overdosed.

When it comes to addiction problems, then, we will have to take a very close look at the pharmacology of those substances and the reasons why somebody self-medicates with them to overcome some kind of problem. Also, we will have to look into genetic predisposition to determine how vulnerable individuals are and how they could be advised to steer clear of certain behaviors or substances. Only then will we be able to come up with successful therapy options. Furthermore, we should not stigmatize or outlaw those who are productively using various substances to enhance their lives, as long as they do not harm themselves or others.

Cannabis and the class of cannabinoids are very promising when it comes to treating people afflicted with alcohol, opioid, methadone addictions, as several studies have shown.[138] It is surely a long way to finding out more about which cannabinoid can be most effectively used for various substance abuse addictions, but we have already begun to understand that cannabinoids may be helping people with substance abuse disorders. Future medical studies will hopefully bring even more evidence on successful therapeutic strategies using cannabis for addictions such as substance abuse disorders.

Enhancement Uses and Addictions

Sadly, though, the scientific community has virtually no understanding still of the mental enhancements of cannabis use described in this book which can powerfully contribute to fighting addictions and stimulate personal growth.

This lack of understanding leads to two kinds of problems: first, many enhancement users will be classified as "drug abusers" – because for many, consuming cannabis is always *per se* "drug abuse" of an addictive drug. If we do not understand that there are many users out there who are using substances like cannabis, MDMA, or other legal and illegal substances to substantially enhance their minds, then we will not come to a full understanding of the motivation and nature of this use, with harsh consequences for individuals, families, and society at large. Second, if we do not understand how cannabis can enhance our mind, we cannot understand in which ways an intelligent and informed use of cannabis (or other substances like LSD) may contribute to fighting addictive behaviors.

Some years ago I was talking to a middle-aged man at a party and mentioned my cannabis use and research. He looked at me slightly shocked, and said, fully convinced of himself: "I am not using drugs!" I couldn't help giving him a mild smile, looking at the cigarette in his left hand, which also held a glass of Vodka Red Bull. Altogether, he was holding a cigarette containing dozens of synthetic and natural psychoactive substances, and a drink combining sugar, alcohol, caffeine, Taurin, and nicotine. He was still completely oblivious to the fact that he was under the influence of several highly addictive psychoactive substances. Advertising and decades of disinformation campaigns had successfully managed to convince him and so many others that there are "alcohol, tobacco, and *drugs*"- a classification scheme presupposing that alcohol and tobacco are something other than a drug.

For a few long seconds, I just smiled at him quietly and the many psychoactive substances in his hand. He still did not get it. And then I had an idea. I would just write an essay. This essay.

And, yes, admittedly, when I wrote the essay, I used the high, and many of the thoughts expressed here originated in high thinking.

(22) Close-up of the upper part of a White Haze flower.

X. Why the Prohibition Against Cannabis Has to End

A scientific truth does not triumph by convincing its opponents and making them see the light, but rather because its opponents eventually die and a new generation grows up that is familiar with it.

Max Planck, 1858-1947, physicist,
founder of quantum physics

Watanuga Lahele is radiating. His glassy eyes peep out under a large, conical straw hat, his movements slightly erratic. He has been chewing on a dark-greenish Kalangi root and the psychoactive substance tetralin it contains now clearly shows its euphoric and mind-altering effects. Lahele sits at a huge wooden table, like every year on May 1st, at the Kalangi root festival in Bomaki, the capital of the Republic of West Africa. Millions of visitors have come together to celebrate and to get collectively intoxicated at the festival. Lahele does not quite manage to get up from the table; he stumbles and falls sideways onto some other visitors. Soon, several people get in a brawl. The Kalangi root often leads to disinhibition and makes its consumers more aggressive. In the Republic of West Africa, tetralin is completely legal, despite its strong mind-altering effects and various dangerous side effects.

A Legal Drug Festival in Africa

During the Kalangi festival, there are dozens of rapes every year committed by intoxicated visitors, and sexual harassment is quite common. Many visitors end up in emergency rooms after dangerously overdosing on the drug. These aspects are only the tip of the iceberg when it comes to the destructive potential of the psychoactive substance tetralin for individuals as for society. Its effects on human perception and motor control during strong intoxication are so profound that thousands of consumers in the Republic of West Africa get injured or die in more than 13 thousand car accidents every year. More than 1.5 million men and women between 18 and 65 abuse tetralin despite its negative physical, psychological, and social consequences. In a population of roughly 80 million people, more than 70 thousand die every year from the long-term side effects of using the drug, often combined with tobacco consumption. The side effects of chronic abuse of tetralin on the human body and mind are disastrous. Prolonged heavy use can cause high blood pressure, liver cirrhosis, gastritis, pancreatitis, nerve cell damage, psychosis, laryngeal, liver, stomach, or pancreatic cancers, as well as cardiac insufficiency, depression and suicide, anxiety disorders, impotence and a diminished capacity for sexual experience, forms of dementia such as Korsakov's syndrome, a pitiful state in which patients forget earlier experiences and are also incapable of remembering new experiences; their personalities simply fall apart.

Many of the side effects of long-term use are irreversible. According to a study by English scientists published in the renowned medical journal *The Lancet*, tetralin appears to be less toxic than hard drugs like crack or heroin, but ultimately more dangerous if we consider the catastrophic sociological consequences. The abuse of tetralin destroys tens of thousands of families each year. According to estimates from government reports, almost every third rape happens under the influence of the drug. It is also involved in approximately one-third of cases of aggravated assault and manslaughter.

How is it possible that the Republic of West Africa does not prohibit this dangerous substance, that they even celebrate and worship it in their society? Why is this culture so much more permissive than other countries, where a substance like cannabis, which according to many experts is even much less dangerous than alcohol or nicotine, is strictly prohibited?

These are the wrong questions, of course. Some better questions would be: Have you realized that I was playing a game with you, that the Republic of West Africa does not exist, and that there are no such things as the 'Kalangi root' or the drug 'tetralin'? Do you have at least a suspicion of what I was talking about?

The Largest Drug Festival in the World

Please visualize again the scenery of the drug-consuming West African at the festival. From there, let's cross-fade to reality. We are now in the city of Munich, Germany. Replace Watanuga Lahele with Michael Müller, the 33-year old project manager of a car company. He wears his traditional Bavarian festive costume: leather pants, a blue and white checkered shirt, and a big grey conical felt hat. Michael, a rather timid guy, usually, and a diligent employee, is sitting in a giant tent and belting out Bavarian folk songs while occasionally grabbing the backsides of passing waitresses and making lewd comments.

He is drinking his fifth strong Bavarian festival beer, and his perception and motor coordination have deteriorated just as described for Watanuga Lahele above. He is attending the biggest and most famous drug festival in the world, the *Oktoberfest* in Munich, in Germany, central Europe, the self-proclaimed epicenter of civilized European post-Enlightenment rational society. Approximately 6 million visitors from all over the world visit this festival every year.

So, no, Watanuga Lahele does not exist, the *Kalangi* root does not exist, and there is no drug called *tetralin*. The Bavarian Oktoberfest famously exists, alcohol exists, and all the statistics and facts cited above given for *tetralin* actually roughly apply to the Oktoberfest and alcohol consumption in Germany or England in the last years. The statistics given above on fatalities, rapes under the influence of *tetralin*, and the long list of horrible side effects, all these statistics are well-known facts about alcohol consumption, and it certainly does not look significantly different for many other countries. And yes, in fact, the mentioned study in *The Lancet* really does say that alcohol should be considered more dangerous than crack or heroin if you consider the destructive social consequences of alcohol.[139] The study was published by Prof. David Nutt, the renowned expert and chief drug advisor of the British government, who was fired after publicly stating what many colleagues in his field believe: that substances like cannabis, LSD, and ecstasy are, from a pharmacological point of view, far less dangerous than alcohol.

So, do we need a strict prohibition against alcohol to protect our society and especially our youth from this dangerous drug? We could punish even the possession of a small amount of alcohol and expel students from high school when consuming any small amount of alcohol – how about a zero-tolerance policy? Illegal alcohol drinkers, which we would then call "criminal drug consumers", would be stigmatized for the abuse of even tiny amounts of alcohol, such as for drinking half a glass of beer, and, if detected, they would get an entry into their criminal records. We could force them to take unannounced blood tests to get a job and keep it.

Certainly, we would also have to start repression of epic proportions to fight the illegal market created by our prohibition, with myriads of policemen, undercover agents, soldiers, and border control troops to go after producers, dealers, smugglers, and consumers. We would have to expect a huge illegal market with millions of criminal users, so we would have to send millions of users to jail.

Of course, we now know that alcohol prohibition was a really bad idea. But, as a reaction to the story of Watanuga Lahele, a prohibition may have seemed to some readers as a reasonable measure to fight the use of the Kalangi root – a drug for which I described in detail the side-effects known for alcohol – and those effects only. What is wrong with our perception here?

What about our racial and cultural prejudices about African societies? Did they play a role when you read the story of Watanuga Lahele? How did you feel about this African society? Maybe you felt that it is too primitive to deal with the challenge of sensibly regulating a dangerous drug? I leave it to the reader to reflect on this.

Why do we still see so many nations with a strong cannabis prohibition, while a much more dangerous drug like alcohol is legal in most countries despite all of its destructive toxicological side effects and its negative impact on society?

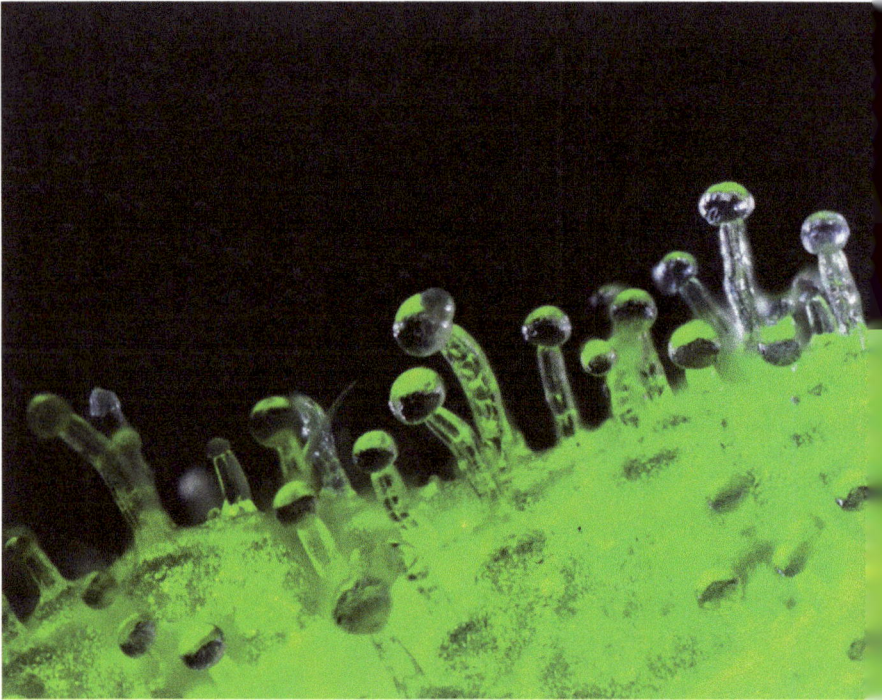

(23) Trichomes on a translucent leaf on a Mexican Sativa flower. Evolutionary biological theories about the function of trichomes and the production of cannabinoids assume they have protective functions. First, the trichomes form a protective layer against insects, but also protect against UV-B sunlight, wind, and low humidity. Another function could be the nematocidal and antibiotic effect, as well as the bitter taste of cannabinoids to deter herbivores.

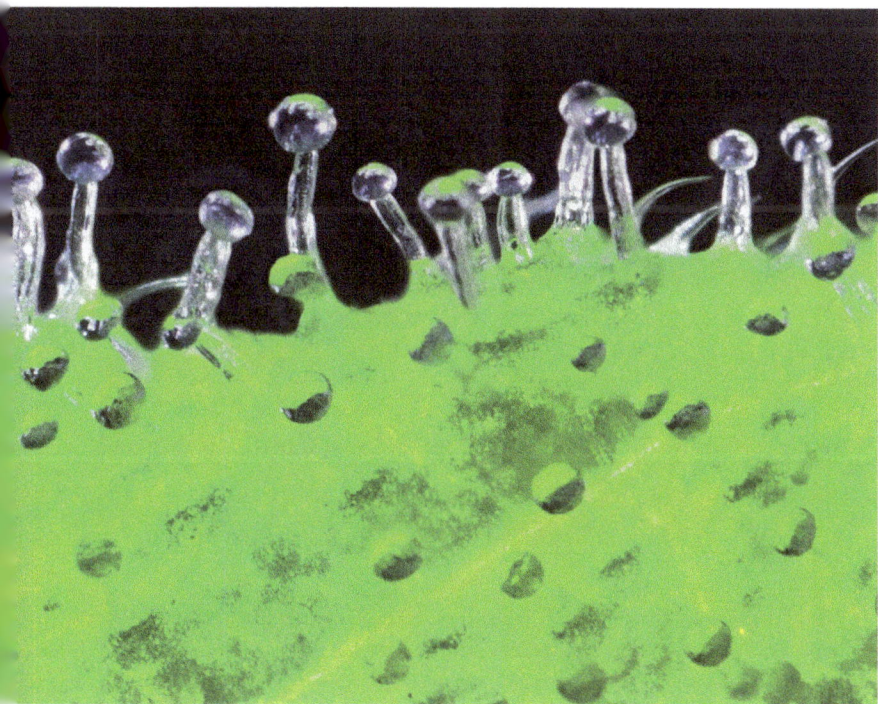

The Prohibition of Cannabis in the U.S.

The prohibition in the U.S. still costs billions of dollars and creates a gigantic black market and criminal structures. In the last years, however, many governments around the world have reformed their laws on medical cannabis use, and many others have legalized cannabis for responsible adult use. *Pubmed*, the research database of the American National Institute of Health, shows almost 12,000 results of studies and articles on the human endocannabinoid system. Many countries all around the world have made very good experiences with medical marijuana laws and programs. Medications based on cannabinoids have been clinically proven to be safe and effective, and there is more than enough data from studies with natural cannabis products showing how effective cannabis can be for the treatment of chronic pain, epilepsy, and many other medical conditions. [140]

Yet, in 2022, unbelievably, cannabis is still classified as a Schedule 1 substance under the CSA (controlled substances act of 1970) in the U.S., which means it is classified as not safe, having a high potential of abuse, and not accepted for medical use. Schedule 1 drugs cannot be prescribed and it is hard for researchers to investigate them, as they need special permission from Federal agencies.

Why is it still such a hard fight for so many countries including the U.S. to change the law and set reasonable regulations for medical cannabis use and adult use in place?

We have to take a look at the history of cannabis prohibition to understand where we are now. The full story of the cannabis prohibition in the U.S. and the rest of the world is complex and multifactorial. Yet, I'd like to highlight at least some aspects of it here to show how irrational the decisions of lawmakers have been from the very beginning – and why we are still seeing highly irrational forces at play.

In the 1920s, smoking marijuana was only common for some minority groups in the U.S.; specifically, blacks, particularly in the jazz scene, and some Mexican immigrants.[141] The prohibition against marijuana had started at the beginning of the 19th century, presumably for various reasons. The recovering alcohol industry probably wasn't too fond of a cheap intoxicant.[142] For a long time, historians believed that the cannabis prohibition beginning in various Southern U.S. states around the 1910s and '20s was motivated to a large degree because of a need to target and criminalize cannabis using Mexican immigrants, who had been lured to the Southern states as a cheap labor force but were not needed anymore at a time of economic depression.[143] According to researchers Richard Bonnie and Charles Whitebread, these Mexican laborers could be easily targeted and singled out for their marijuana use. It is questionable, whether this "Mexican hypothesis " really depicts the main cause for the beginning of the prohibition in the U.S., even though it may have been one of the driving forces.[144]

Still, historical evidence strongly suggests that xenophobic stereotypes about Mexican workers and their alleged criminal and aggressive behavior played a role in the early days of prohibition and were later used to fuel a long and influential disinformation campaign to support the prohibition of cannabis.

From its early days on, cannabis prohibition in the U.S. was reinforced with a disinformation campaign that began in the 1930s. After the end of the catastrophically failed alcohol prohibition, the federal prohibition agency had to look out for a new task to stay in business. To that end, Harry G. Anslinger, the head of the new Federal Bureau of Narcotics (FBN), started a mixed media campaign against *marijuana,* the horrible drug from *Mexico.* It was an important and very effective propaganda decision to use the Mexican slang term "marijuana" and label it as a Mexican drug. Most of the U.S. population did not know that marijuana was cannabis, a substance that had already been in use for decades as a medicine for various purposes and was freely available in drug stores all over the U.S. Horror stories about marijuana were invented and placed in newspapers nationwide and found their way into radio spots and propaganda films. Anslinger let his spin-doctors portray marijuana as a deadly poisonous drug that would turn consumers into aggressive and out-of-control rapists and killers, leading even the occasional user to chronic insanity and death. One of his extremely cynical strategies was to exploit existing racial prejudices against those minority groups who predominately used marijuana at the time: Mexican immigrants and black jazz musicians like Louis Armstrong.

"To gain support for his crusade, Anslinger depicted marijuana as a sinister substance that made Mexican and African American men lust after white women. One of the worst things about marijuana, according to the FBN chief, was that it promoted sexual contact across color lines. 'Marijuana causes white woman to seek sexual relations with Negroes,' Anslinger frothed. He rang alarm bells in segregated America, warning that black and whites were dancing cheek-to-cheek in tea houses and night clubs, where pot-maddened jazz bands performed what the Hearst papers called "voodoo-satanic music."[145]

Anslinger's scare tactics turned out to be remarkably effective in a time when the "Jim Crow" racial discrimination and segregation laws were still in place, laws designed to humiliate "persons of color." In the 19th century, "Jim Crow" ("Jim, the crow") had been a racist stereotype of a black man as lazy, singing, and often stealing, and who was often portrayed in Minstrel shows. [146]

These laws made it impossible for blacks to vote, and to sit on juries, and they were segregated from others in all areas of modern culture, in schools, restaurants, or hospitals. Black citizens were expected to show deference to "white people," but clear rules of conduct were usually not given, so it was more or less arbitrary for a white person to feel offended. From 1883 to 1941, more than 3200 black citizens were lynched, with rape as the second most common accusation after murder or attempted murder. [147] Hangings became better known because some of the victims were photographed and postcards showing them were circulated as souvenirs; but many others were shot, burned alive, thrown off a bridge, or dragged behind a car. The Jim Crow laws were only fully abolished by the Civil Rights Act of 1964. Only if we remember how strong racism was embedded in the American culture and the laws of the 1930s do we understand how sinister Anslinger's campaign against cannabis really was.

Anslinger's disinformation campaign eventually led to the prohibition of marijuana in the U.S. Medical experts of the American Medical Association testified against its prohibition but were ignored.[148] The House of Representatives followed Congress and approved the "Marijuana Tax Act", which effectively started the prohibition in 1937 after a 90-second long discussion of the bill. During the said discussion, two questions were raised. In response to the question about the opinion of the American Medical Association (AMA), a committee member falsely informed them that the AMA would fully endorse this measure. The other question was to summarize the purpose of the bill: The speaker of the committee replied: *"I don't know. It has something to do with a thing called marijuana. It is a narcotic of some kind".*[149]

It is hard to believe that such an extensive prohibition would have its roots in a disinformation campaign based on manufactured racist lies and purely invented claims about the toxicological effects of cannabis. Yet, the history of marijuana would take an even more bizarre turn. In the following years, more and more experts doubted Anslinger's claims. The first serious interdisciplinary study on the subject of marijuana, initiated by New York's Mayor Fiorella LaGuardia in 1944, refuted almost all of Anslinger's claims.[150] Experts found that cannabis use would not lead to violent or criminal behavior, as Anslinger's propaganda had infamously claimed, that it would not act as a gateway drug, and they could not find scientific evidence for the claim that marijuana would be dangerously physically harmful to users.

Faced with this evidence, the outraged Anslinger struck back with a new strategy – another disinformation campaign geared to the new political climate. In the heated atmosphere of the anti-communist McCarthy era, Anslinger now claimed that marijuana had to be outlawed because it would make users too peaceful and because communist China would illegally bring marijuana to the U.S. to undermine the defensive morale of the military and the public.

"Scarcely a decade after telling Congress that cannabis was 'the most violence-causing drug in the history of mankind', Anslinger did a 180-degree switcheroo and declared marijuana a threat because smoking the herb turned people into docile zombies – to the point where stoned U.S. citizens would be neither willing nor able to fight the Red Menace within and without."[151]

The absurd twist in Anslinger's argumentation did not seem to bother politicians much. The prohibition was prolonged and the penalties would get even more severe under the "Boggs Act", which was signed by President Truman in 1951.

Other economic factors may have played an important role in the initial and later motivation for the ongoing cannabis prohibition, such as the interests of the pharmaceutical industry, the tobacco and alcohol industry, the paper industry including the media empire of William Randolph Hearst, and the chemical industry, all of which had a vested interest to keep cannabis prohibited. However, it is still unclear and historically less well discovered to what extent those industries and Anslinger's connections to them played a role in the dynamics of his prohibition.

Anslinger was not only the chief commissioner of the Federal Bureau of Narcotics from 1930 to 1962. Additionally, he served as the American representative of the United Nations Drug Commission and used his influence to implement the "Single Convention on Narcotic Drugs", in which marijuana was put on the same level as opiates. The global ban of cannabis cultivation can primarily be traced back to the efforts of a man whose main sources of information were publications of the tabloid press, as a posthumous searching of his files reveals. Anslinger's disinformation campaign led to a lasting cannabis prohibition and effectively shaped the legislation as well as the public's image of cannabis worldwide. Even today, our understanding of cannabis is strongly affected by negative associations drilled into the public's consciousness by Anslinger's previous disinformation campaigns.

(24) The mature stigmas of a Mexican Sativa flower. The calyx at the bottom left, from which they arise, is clearly visible, as are the small mushroom-shaped trichomes. The cannabinoids, terpenes, and flavonoids, many of which are still largely unexplored, are found mainly in the trichomes, calyx and stigmas, but also in smaller amounts in the leaves and other parts of the plant.

Substance Prohibitions as Part of a Multi-Faceted Power Struggle

In the history of humankind, substance prohibitions have always been motivated by various factors, many of which have nothing to do with their real toxicity and an honest attempt to protect the citizenry.

To understand what factors come to play we need to look at the bigger picture of the nature and dynamics of other substance prohibitions in history. If we take a closer look at historical substance prohibitions, we find that many of them are not mainly motivated by a government or a ruling class trying to protect the health of its citizens, but by a ruling class trying to exploit or oppress a certain ethnical, religious, political, or "racial" group. Economic motivations play a big role, such as economic protectionism, as well as religious beliefs, and, also, false assumptions about the real effects of a substance.

Let us take a short look at two historical coffee prohibitions. During the coffee prohibition under Murad IV, Sultan of the Ottoman empire from 1623 to 1640, coffee drinkers were threatened with beating and the death penalty, and many coffee consumers were executed. The justifications given for this prohibition were, among others, religious reasons, arguing that coffee would intoxicate drinkers in a way that was forbidden to Muslims, and medical reasons, claiming that coffee was bad for the human body.[152] The real reasons for Murad IV to crack down on coffee houses were probably different:

> "In his childhood, explains Ottoman political historian Baki Tezcan, his brother Osman II was deposed and brutally murdered by the janissaries, a military group that had grown increasingly autonomous and

discontented. (...). Murad IV knew that demobilized or under-employed janissaries frequented coffee shops – and used them to plot coups. Some coffeehouses eventually used janissary troupe insignia as their signage. Murad IV was also likely aware, argues Ottoman historian Emingül Karababa, that a conservative religious movement, which opposed Sufis and social innovations connected to them, including coffeehouses, was rising in his empire. "He did not want to have tension and uprisings in the society under his rule," as he sought to wrest control from the janissaries, argues Karababa. So it was doubly in his interest to oppose coffeehouses. [153]

In Prussia, coffee drinking was prohibited by Frederick the Great as of 1766, while he encouraged the use of beer, which, he stated, would be healthier than coffee. He may have believed that coffee drinking could cause infertility, as many Prussian doctors did at the time, but that was hardly a reason for him to abstain from drinking it himself. The coffee prohibition was motivated by the King's need for money after a long war and because he was convinced the money spent by so many ordinary people on coffee would get drained from his economy and go to the coffee-producing Ottoman Empire. [154]

German beer brewers had already started to complain about sales losses under the pressure of the new competition. So, Frederick the Great, who would famously often drink his coffee spiced with pepper or mustard, sent out coffee sniffers, retired French soldiers hired for big salaries. They made their rounds stopping ordinary men and women on the street – not the nobility or the upper class, of course – to sniff on them. They searched their houses without a warrant to seize the now illegal psychoactive substance and to punish the convicts. According to rumors, many of these soldiers, then, sold the seized coffee on the black market. Frederick the Great imposed luxury import taxes on coffee to support his prohibition, and set in place a monopoly on the coffee trade which made him rich. All this did not show any visible effect on the consumption habits of his people, which made it more expensive for ordinary people to get coffee, but the nobility and the rich were of course still able to afford it. Cheap black market coffee roasts came over the border in hay stacks, coal charges, and even coffins, and market ladies had breast bandages designed to get their black market coffee to their customers. [155] The coffee prohibition was only ended by his successor – probably because of the insight that the monopoly may have made Fredrich the Great rich, but the Prussian state had lost a lot of money because of the huge black market. [156]

As we can see, then, in these two coffee prohibitions, socio-economic factors and power struggles play a major role. These struggles brought with them the need to control or repress certain groups of people and played a much bigger role in the imposing of prohibitions than alleged medical or religious reasons which were publicly given as justification. This is an important observation: Substance prohibitions in history have often been justified with the alleged toxicology of a substance, or its effects promoting behavior that would clash with religious or moral rules, but the real driving forces behind these prohibitions were different. It was of course easier to mislead those to be controlled and punished by tricking them into believing a justification alluding to a moral framework they already accepted. Historically, we can say that in some prohibitions, moral entrepreneurism certainly also played a role in substance prohibitions, as they were perceived to promote or cause behavior that would conflict with religious or moral values.[157] Importantly, however, from a contemporary perspective, we would re-evaluate those prohibitions generally as motivated by reasons that are not in the interest of consumers. Also, we can see that past prohibitions often were ineffective, which concerns reducing the amount of consumption of a certain substance despite strict reinforcement, leading to large black markets with huge losses for the civil society because of tax evasion.

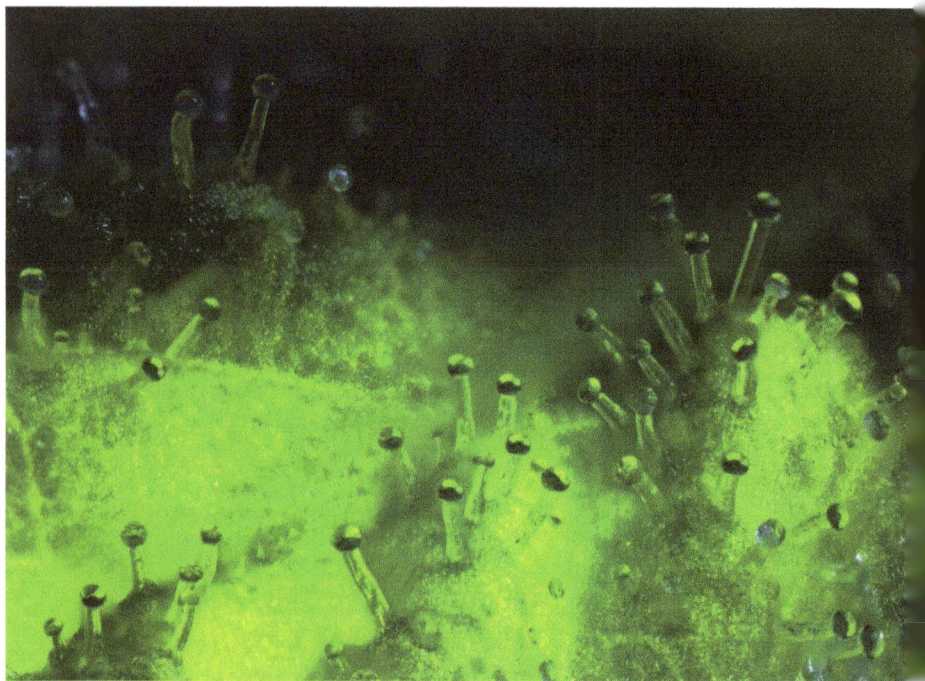

(25) Another "landscape" with trichomes on a flowering The Pure plant. As humans have been cultivating and breeding cannabis for thousands of years for various purposes, including for the high, ultimately the psychoactive and medical effects of their biochemicals have provided one of their greatest evolutionary advantages in the last few thousand years, greatly influencing the development and spreading of countless cannabis varieties over the past millennia.

The Ongoing Disinformation-Campaign
Against Cannabis

So, past prohibitions were driven by irrational forces.
But what about today, or the last few decades? We have
changed, right? We cannot take our criticism of past
substance prohibitions as being irrational and apply it
to modern prohibitions, for instance the prohibition of
cannabis. After all, we are now living in a society in an
ongoing Age of Enlightenment, a time in history with huge
progress in the empirical sciences and a society that is
mostly informed by it, following ideals of rational thinking,
liberty, progress, tolerance, equality, and the separation of
church and state, thus limiting the influence of irrational
belief systems. If we now have a prohibition, it cannot
crudely paint a portrait of the toxicology of a substance
and still be in place because of irrational racist beliefs,
economic interests of certain groups, lobbying efforts,
or governments who behave in opposition to what their
scientists say. Right?

When I published the first version of this essay in
Germany almost 10 years ago, I felt like there were still
many naïve intellectuals out there who would simply
rely on the current cannabis prohibition out of their basic
reliance on our government(s) as acting more or less
rationally, all in all, and *"doing the right thing."* I guess
things have changed, dramatically, and I do not know if
I even have to address this kind of naïve view anymore.
Since 2013, several democratic elections have been
undermined by right wing groups, with vicious targeting
and mass manipulation techniques aided by Facebook and
Cambridge Analytica.[158] Donald Trump became elected
and lied to his public almost on an hourly basis, and then

tried to motivate his fan group to stage a coup when he did not become re-elected.

Some years ago, the EU almost fell apart under the pressure of right-wing and authoritarian movements, fueled by vicious social media campaigns, and those ultra-nationalist and right-wing forces are gaining more force all over the world. In 2018, the Chinese dictator Xí Jinpíng made sure he can stay in charge until he dies, basing his reign on censorship, mass surveillance, and a personality cult. Hundreds of thousands of Uighurs and many other Muslim minority groups have been subjected to mass internment and torture in China, and millions of Muslims are under systematized mass surveillance.

In 2021 Agnès Callamard, Amnesty International's Secretary-General stated:

> *"The Chinese authorities have created a dystopian hellscape on a staggering scale in Xinjiang. Uyghurs, Kazakhs, and other Muslim minorities face crimes against humanity and other serious human rights violations that threaten to erase their religious and cultural identities. (...) It should shock the conscience of humanity that massive numbers of people have been subjected to brainwashing, torture, and other degrading treatment in internment camps, while millions more live in fear amid a vast surveillance apparatus."*

Russia's President Putin, who recently also made sure that he can be dictator of Russia for his lifetime, just waged a war on Ukraine, targeting hospitals and killing civilians and refuges. As I am writing this article, images and articles about civilians and journalists killed by the Russian Army in the Ukraine pop up in the background. Some days ago, Putin made sure to pass a law punishing those who call his war a war (and not a "military operation") with up to 15 years in jail.

In general, then, I guess we are all not that naïve anymore when it comes to believing in the triumphal march of science and rational thinking in our governments and society. Still, as far as I can see, many of those who believe in the effectiveness of the cannabis prohibition think that we are well informed by science to uphold the prohibition to protect our children and others.

So, to see why the current prohibition of cannabis is still ongoing, let us distinguish two types of irrational factors which keep the prohibition rolling. Importantly, there are long shadows from the past which come from long disinformation campaigns and the myths that have been drilled into our minds for decades. As we will see, misconceptions are now deeply embedded in our language and thinking about cannabis and other psychoactive substances, and they still cause irrational reactions. Second, we need to look at the current irrational dynamics of politics to see why politicians and governments need such a long time to rationally adjust their policies to our current knowledge – or better, to what we could have known for decades now, if we would have listened better to scientists.

Let me first look at how disinformation campaigns changed our language and, thereby, our thinking about cannabis. Take a look again at Anslinger's diabolically smart decision – "smart" from a propaganda perspective – to use the notion of "marijuana" for his campaign. Anslinger and the yellow press he cooperated with used the Mexican slang term "marihuana" to draw on strong racial and cultural prejudices against Mexicans and blacks. This strategy enabled him to exploit the fact that linguistically speaking, white middle-class citizens would not understand that marijuana was cannabis, a substance they already knew and valued since the mid-19th century, when cannabis tinctures were introduced as a medicine to the U.S.[159] Those who knew cannabis from the pharmacy did just not connect the dots.

Anslinger's disinformation strategy has been refined by later prohibitionists. The new PR strategies can most prominently be made visible by analyzing the deliberately misleading rhetoric of the slogan "war on drugs," designed and introduced during President Nixon's presidency. The slogan and the worldview it subconsciously suggests are wrong in many ways. To illustrate, let us ask a few questions: If we need to fight a "war on drugs", then all drugs are evil and we need to fight them? Really? How about aspirin and other pharmaceutical drugs? How about alcohol and tobacco? How about the drug coffee? Furthermore: Is it a "war on drugs" – or, mainly, on a specific group of drug consumers? Who are the victims? The metaphor "war on drugs" suggests that this is a conflict on eye level. It's a war between law enforcers against drugs – specifically, cannabis plants armed with their deadly poison. Accordingly: victims on both sides would be law enforcers – and ... "drugs". And why be concerned about this war on cannabis – the victims would just be dead cannabis plants. Drug producers, users, dealers, or patients using drugs or self-medicating with drugs, all do not enter the picture here. It is just a war on drugs, right?

As the linguist George Lakoff showed us, we need to understand that slogans express metaphors, and these metaphors express certain worldviews.[160] If you argue against the "war on drugs", then, still, by your very use of the slogan you support a worldview that you probably do not have, namely, that this repression is really about all drugs, or mainly about their evil toxicology. This suggests that we do not need to care for or help traumatized people using medical and other "drugs" because the problem is in the "evilness" of the drugs. So, if you argue that we should end the "war on drugs", you say we should all surrender to their toxicity.[161]

These metaphors are still strong and we keep on using them. Thus, we carry on the worldview designed by Nixon's PR strategists. But did the Nixon administration want to fight drugs? In a 1994 interview published only in 2016, Nixon's top advisor claimed:

> *"The Nixon campaign in 1968, and the Nixon White House after that, had two enemies: the antiwar left and black people. You understand what I'm saying? We knew we couldn't make it illegal to be either against the war or black, but by getting the public to associate the hippies with marijuana and blacks with heroin, and then criminalizing both heavily, we could disrupt those communities. We could arrest their leaders, raid their homes, break up their meetings, and vilify them night after night on the evening news. Did we know we were lying about the drugs? Of course, we did."*[162]

Ehrlichmann certainly was bitter about Nixon when he made his statement. So, was he just cynical and bad-mouthing Nixon? Probably not. Given many of Nixon's well-known racist statements and his contempt for the rising counter-culture of the '60s and the civil rights movement, Ehrlichmann's statement probably was not that far off target.

So, sadly, we know that Nixon was a liar and a racist, but we are still using slogans today which entail a worldview designed by Nixon's aides. Even if we are against the "war on drugs", we suggest his worldview and allude to its false presumptions.

Let me only point to one more example where a learned use of language can lead to a subconscious acceptance of a worldview that comes far from the past. Imagine you take your children to the Zoo, as they grow up, and you tell them there they will see "animals, and kangaroos". Would you be surprised to find out at some point that your kids are convinced that kangaroos are not animals? Well, from their perspective, if they are animals, why did you not subsume them under the term?

Now, what if you hear and read about all those people who used alcohol and drugs? Alcohol must be something different than a drug, right? So, calling the Octoberfest the largest drug festival in the world sounds . . . weird, no? Why? If you think about it, you know that alcohol is a drug, no?

We have all been linguistically conditioned to some degree, and this mechanism is still so strong, subconsciously, that even convinced advocates of cannabis legalization use phrases like "alcohol and drugs".

Cognitive scientists speak of "cognitive priming" if a previous process of conditioning leads to certain associations, such as when we hear a notion and automatically associate images or have certain expectations previously paired with this notion. The ongoing prohibition keeps generating the stereotypes which it needs to survive: cannabis consumers are made criminal by prohibition laws and so we associate cannabis use and everything having to do with cannabis with criminal behavior.

Additionally to these toxic linguistic remnants of the past, countless myths about marijuana continue to circulate today, even though they have long been refuted by scientific evidence. In their study "Marijuana Myths, Marijuana Facts", the authors Lynn Zimmer and John P. Morgan collected all the myths and quotes in detail and summarized the scientific studies debunking those myths.[163] To name only a few, they investigate and debunk the myths that marijuana "has no medical value," that it is a "gateway drug to harder drugs," that it "causes an amotivational syndrome," that it "kills brain cells," "causes crime," or that it is "more damaging to the lungs than tobacco".

But, again, at this point, modern prohibitionists probably tend to believe that our contemporary approach to regulating and prohibiting cannabis is better informed by science today and, therefore, justified. Irrational forces like those in the past can hardly play such a big role today, right? Well, first of all, I would argue that politicians fall prey to the same myths and toxic linguistic remnants of the past. Yet, furthermore, we can see that there are additional irrational dynamics in politics at play that add to the malignity of the cannabis prohibition.

Let's look at what highly acclaimed pharmacologist Prof. David Nutt, the former drug advisor for the British Government, says in his new book about how rational the decision-making of politicians is when it comes to regulating psychotropic substances. Around 2007, Nutt reviewed the scientific evidence on many substances including cannabis, and suggested rescheduling cannabis to a much less harmful substance class than it had been in before. Ultimately, Nutt was sacked by his government in 2009 despite huge protests from the scientific community.

He later summarized the political process in a very open statement in his recent book:

"With hindsight, I think both prime ministers wanted to have cannabis reclassified primarily to keep the

votes of the majority of the public. Most people accepted the dominant dogma of the time, that drug harms could be reduced by drug users being punished. It follows that in order for a Labour government to keep the support of those politically in the centre, they had to be seen to be as 'tough' on drugs as the Tories. (...) In fact, the government did likely believe there were votes in drugs – but in punishing people for using drugs. Previously, the police had been relatively relaxed about cannabis because 'stoners' didn't cause them nearly as much trouble as drunks. Now, as one of the criminal justice reforms, police were given performance targets, and the easiest way to achieve them was with the newly introduced cannabis street caution that the police could use to deal with people caught with cannabis for personal possession.

The number of cannabis offenses soared. And the policy – or its implementation – was clearly racist. Three to four times more black and ethnic minority people were prosecuted than cannabis use statistics would have predicted.

This campaign led to over a million young people – mostly men – being given a criminal record and so created an underclass who found themselves faced with severely limited job prospects. "[164]

Nutt, of course, is not alone with his experiences confronting politicians as a scientist. The sad historic record is that government agencies of various countries repeatedly issued comprehensive scientific studies on its individual effects on users and their impact on society, only to have the governments ignore the results. All the most important comprehensive studies issued by governments on the dangers of cannabis use refuted

Anslinger's myths about the dangers of cannabis and recommended decriminalization, whether it was the *Report of the Indian Hemp Commission* (England 1894), the study of the *Mayor LaGuardia Commission* (1944), the *Baroness Wooten-Report* (England 1968), the *Report of the Lading-Commission* (Canada 1972), the *Shafer Commission Report* (1972), or the *Report of the Canadian Government Commission of Inquiry Into the Non-Medical Use of Drugs (Canada, 2002)* – to name only some.

The Destructive Consequences of the Prohibition

Even if we agree with a more scientifically informed view about marijuana, which clearly states that the overall risk potential of cannabis is much lower than that of a legal drug like alcohol, we certainly have to take the risks seriously. Every psychoactive substance brings risks for consumers. Adolescents are especially vulnerable to problems with cannabis abuse – mainly not because it would physically harm them, but because for some, chronic and uneducated use may lead to a form of escapism which can heavily contribute to failure at a critical stage in their educational careers.[165] Adolescents and people with personal instabilities need the special protection of our society. Tragically, though, it is exactly in this respect that the prohibition of cannabis fails the most. In recent surveys, high school students say that despite the strict and ongoing prohibition, cannabis is easy to get on the street these days. [166]

Despite all the efforts, prohibition is not only ineffective, but it is also destructive and deadly on a monstrous scale; parents lose their jobs or even go to jail for even minor cases of possession, and students are expelled from school or college. Since 1995, there have been more than 17 million cannabis-related arrests in the U.S. alone. Families are torn apart and the prohibition generates a huge illegal market, in which criminal organizations violently fight for control and consumers often get laced and contaminated marijuana. And this is only the beginning. Worldwide, the consequences of the illegal drug trade have generated a huge death toll and we know that regulating cannabis and other substances instead of prohibiting can greatly diminish the power of drug cartels and their detrimental influence on society.[167]

Even worse, citizens, and especially our youth, loses their trust and respect for a government's irrational prohibition of a substance widely known to be medically highly valuable and far less dangerous than legal substances such as alcohol or tobacco. This loss of trust is disastrous when it comes to the general relationship between citizens and their state. In connection with the behavior toward drugs, it may also lead many teenagers to underestimate the much more dangerous toxicity of legal drugs like alcohol. The ongoing irrational treatment of cannabis makes it impossible to correctly and objectively educate an entire generation of adolescents. As a result, we miss our chance to present ourselves as a reliable and rational source of information to effectively educate millions of citizens and generate a more healthy attitude and behavior towards psychoactive substances.

We can protect our adolescents and citizens much more effectively with a regulation of cannabis similar to that of alcohol or tobacco. For the government to be credible and effective, it has to start behaving rationally – which means, finally, that it should be listening to scientific experts in the field.

(26) Extreme close-up of the trichomes.

The Prohibition and
the Number of Cannabis Users

Prohibitionists often claim that it is an obvious and very intuitive fact that a cannabis prohibition leads to diminishing or containing the number of users, and, therefore, helps to protect our youth and others from existing risks. So, even if we may act irrationally on what concerns our legislation on alcohol, at least we help reduce the risks of cannabis consumption in our population. If we would end the prohibition, they argue, we would send the wrong signal to consumers, and, then, the number of consumers would explode. Well, that is true, intuitively, for some, maybe. Alas, it is only intuitively true for some of those who want to believe in it (which is called an "alternative fact" in the post-Trump era). There is convincing evidence that the cannabis prohibition in several countries is not an effective means to limit or diminish the usage of cannabis. We know that some countries with a strict prohibition like France have higher rates of cannabis users than, for instance, the Netherlands, where the use of cannabis has been decriminalized now for decades.[168]

Also, there are data for Europe showing that ending prohibition sometimes even seems to lead to a decrease in cannabis use, as a study drawing on data from the *European Monitoring Centre for Drugs and Drug Addiction* (*EMCDDA*) claims:

> *"This comprehensive re-analysis of all available data from EMCDDA does not corroborate an impact of changes in cannabis legislation on cannabis use among young people in Europe. Overall, since the 1990s, self-reported use appeared to increase among countries without any policy changes but decrease after both decriminalization and depenalisation of cannabis-related crimes."*[169]

Claims from prohibitionists that the prevalence of cannabis use among younger people would dramatically rise or explode after legalization ("legalization sends the wrong signal!") have been shown to be untrue for the U.S., too.[170]

There is one more aspect in this debate that is hardly ever talked about. If cannabis can be used, positively, for medical purposes, or simply for relaxation, euphoric experiences, stress release, or even for some or all of the mind enhancements described in this book, then there is a meaningful distinction between the use and abuse of cannabis. If cannabis can be seen as a tool, then we can positively use cannabis for therapeutic reasons or to enhance our lives, but we can also abuse cannabis for instance for escapism.[171] In the end, then, the total number of consumers should not be the absolute benchmark of success for our lawmakers. If the number of cannabis consumers goes up, yet many more of them now positively use cannabis instead of abusing it, shouldn't we see this as a success story? As long as there is no rise in health problems connected with this use, and as long as more people are happy and healthy, why would we object to having more cannabis users?

The Rediscovery of the Potential of Cannabis

More than 25 years ago, scientists discovered an *endogenous cannabinoid system* in the human brain, a signaling system in our bodies that uses cannabinoids and their receptors to control several types of physiological and cognitive processes. Since then, scientists have made vast progress in the understanding of this system. This research is already useful to understand what thousands of patients have told us about the medical benefits of marijuana, patients who have claimed that marijuana helped them so effectively in so many ways – without most of the horrible side effects of the prescription drugs they were historically forced to use instead. Of course, the pharmaceutical industry had only limited interest in expensive research of a plant that cannot be patented and which can be grown by anyone in their garden. A substance that can replace many of the existing medications for pain, sleeping disorders, anxiety disorders, depression, inflammation, and countless other medical conditions is not easy to incorporate into a pharmaceutical industry that needs to run on the profit of existing markets authorized to prescribe medications for those conditions. At this point, governments would have to take action and invest money in the medical research of this plant which has been listed as useful in the pharmacopeias of various human cultures for more than 5000 years. Other funds should be used to educate the public about the general risk potential of all psychoactive substances.

It is time now to come to a more realistic view of the subject of psychoactive substances. It is completely unreasonable to think that we should demand complete abstinence from all drugs from our citizens. The use of psychoactive substances has not only been prevalent in all societies throughout human history but goes much further back in evolution. The American psychopharmacologist Ronald K. Siegel, perhaps the world's foremost scientific expert concerning the interaction between animals and psychoactive plants, states:

> *"History shows that we have always used drugs. At every age, in every part of this planet, people have pursued intoxication with plant drugs, alcohol, and other mind-altering substances . . . Almost every species of animal has engaged in the natural pursuit of intoxicants. This behavior has so much force and persistence that it functions as a drive, just like our drives of hunger, thirst, and sex. This "fourth drive" is a natural part of biology, creating the irrepressible demand for drugs. In a sense, the war on drugs is a war against ourselves, a denial of our very nature.* "[172]

Instead of trying to preach abstinence or a *"war on drugs"*, we should educate people who have decided to use psychoactive drugs to come to a more respectful and meaningful relationship with them. To do so, we also need to acknowledge the fact that psychoactive substances have not only risks but also positive potential. The dangers of alcohol are clearly evidenced when we see drunk and aggressive hooligans violently and uninhibitedly beating down their victims. On the other hand, alcohol has a positive potential, not only medically speaking, a potential that many of us have made use of before. The disinhibiting effect of alcohol allows the shy lover to talk to the girl of his heart, the wine gourmet experiences a mind-boggling burst of taste sensations, and writers and many others have used alcohol to better concentrate or facilitate their stream of thoughts. Millions of us use alcohol, not because we are addicted drug consumers, but because we have decided to enhance the quality of our lives despite the risks associated with alcohol use. Many of them urgently need better information about the true risks of alcohol consumption, but we should not outlaw and marginalize them.

The findings in endocannabinoid research and medicine have already begun to show us the huge medical potential of marijuana. They are also beginning to deliver explanations for the thousands of reports of healthy and otherwise law-abiding citizens who use cannabis for inspirational and other life-enhancing purposes. If only a fraction of the myriad user reports about the positive potential of cannabis for various purposes is correct, we must ask ourselves if a prohibition is not a severe intrusion into the personal rights of millions of citizens who have obviously decided to explore that potential for themselves – just as millions of people have decided to use alcohol.

The legalization of cannabis is not a dangerous experiment. The prohibition is the experiment, and it has failed dramatically, with millions of victims all around the world. Cannabis has been used throughout history in many cultures and has been legal for most of this time worldwide. Even in a country like India, where cannabis has been used by large parts of the population for centuries, studies have not found any of the health or other problems that prohibitionists would predict. Users did not run amok in masses and there was no increase in schizophrenia or other mental problems in their population, and similar observations have been made by studies for other countries and times in history. In the Netherlands and in many other countries which decriminalized the use of cannabis there has been no dramatic increase in use and no traffic-related chaos ensued since cannabis was decriminalized.

The Global Commission on Drug Policy, organized to provoke a science-based discussion of humane and effective ways to reduce the harm caused by drugs to people and societies, recently presented its new report, which makes a clear case to end the drug war. Members of this commission include such prominent figures as Kofi Annan, the former United Nations' chief secretary; George Papandreou, former Prime Minister of Greece; Fernando H. Cardoso, the former president of Brazil; former Mexican president Ernesto Zedillo, former Colombian president Cesar Gaviria, Ruth Dreyfuss, former President of Switzerland, Mexican writer Carlos Fuentes, Peruvian writer Mario Vargas Llosa, César Gaviria, former President of Colombia, as well as the former NATO's chief secretary Javier Solana.

Many politicians will have to swallow their pride to rectify the fatal errors of the past. Especially "law and order" hardliners will have a hard time understanding and admitting that their draconic prohibition has never helped to fight criminals, but that it created the catastrophe itself – just as the alcohol prohibition did. We can expect many politicians at best to slowly and silently retreat only under extreme pressure when they gradually come to understand that even with their scare tactics, they will no longer be able to impress voters. Massive lobbying interests such as the pharma, alcohol, and private jail industries still stand in the way of a sensible political change in our drug policies. These lobbyists, also, will need to see that they are out of touch. They need to understand that people have started to wake up from the nightmares Anslinger and his followers made all of us dream.

(27) Trichomes on a Jack Flash plant. In the more than 140 known psychoactive cannabinoids, CBD (cannabidiol) has come increasingly into the focus of scientists in recent years because of its medical effects. Many recent developments in breeding for medically effective hybrids are therefore aimed at a higher proportion of CBD in cannabis plants.

Endnotes

[1] Grand View Research (2022), "Cannabidiol Market Size, Share & Trends Analysis Report By Source Type (Hemp, Marijuana), By Distribution Channel (B2B, B2C), By End-use (Medical, Personal Use), By Region, And Segment Forecasts, 2021 – 2028". *https://www.grandviewresearch. com/industry-analysis/cannabidiol-cbd-market* Accessed March 20, 2022.

[2] Grinspoon, Lester (1971), *Marihuana Reconsidered*. A Bantam Book/Harvard University Press, p. 372-373.

[3] Grinspoon, Lester, *marijuana-uses.com*, Accessed March 19, 2022.

[4] Novak, William (1980). *High Culture: Marijuana in the Lives of Americans*. Massachusetts: The Cannabis Institute of America, Inc.

[5] Tart, Charles T. (1971*), On Being Stoned: A Psychological Study of Marijuana Intoxication*. Palo Alto, Cal.: Science and Behavior Books.

[6] Goode, Erich (1970). *The Marijuana Smokers*. New York: Basic Books.

[7] Lycan, William G. (2019), *On Evidence in Philosophy*, Oxford University Press.

[8] Ginsberg, Allen (1965/2000), "The Great Marijuana Hoax. First Manifesto to End the Bringdown", first published 1965, in: *Deliberate Prose: Selected Essays 1952-1995*, HarperCollins Publishers, New York, p. 95.

[9] Leary popularized these notions, but the idea came from Austrian biologist Ludwig von Bertalanffy in the late 1950s, one of the founders of general systems theory.

[10] Hartogsohn, Ido (2017), "Constructing drug effects: A history of set and setting", *https://doi. org/10.1177%2F2050324516683325*. Accessed March 20, 2022.

[11] Baudelaire, Charles (1966), *Les Paradis artificiels*, Garnier-Flammarion, p. 38.

[12] Tart, Charles T. (1971/2000), *On Being Stoned. A Psychological Study on Marijuana Intoxication*, Iuniverse. com.

[13] Grinspoon, Lester (1971), *Marihuana Reconsidered*, A Bantam Book/Harvard University Press.

[14] Anonymous (2012), "Cannabis and Planetary Surfaces", in: Grinspoon, Lester (ed.), *marijuana-uses.com*. Accessed March 10, 2022.

[15] Amberg, Beth (2012), "Memories of The Moment", in: Grinspoon, Lester (ed.) *marijuana-uses.com*. Accessed March 19, 2022.

[16] Mezzrow, Milton, and Wolfe, Bernard (1946/2009), *Really the Blues*, Souvenir Press, London, p. 72.

[17] Goode, Erich (1971), *The Marijuana Smokers*, Basic Books, p. 164.

[18] Marincolo, Sebastián (2021), *The Art of the High. Your Guide to Using Cannabis for an Outstanding Life.* Self-published, ISBN-13: 978-3981771220.

[19] Baudelaire, Charles, "Poem of Hashish" (1850), translated by Aleister Crowley 1895, https://erowid.org/culture/ characters/baudelaire_charles/baudelaire_charles_poem1. shtml. Accessed March 13, 2022.

[20] Baudelaire, Charles (1966), *Les Paradis Artificiels*, Garnier-Flammarion, p. 47.

[21] Ibid., p. 47.

[22] Galton, Francis (1883), *Inquiries into Human Faculty and Its Development*. Macmillan.

[23] Ludlow, Fitz Hugh (1857), *The Hasheesh Eater: Being Passages from the Life of a Pythagorean*, Harper & Brothers, p. 149.

[24] Tart, Charles T., (1971/2000*), On Being Stoned. A Psychological Study on Marijuana Intoxication*, Iuniverse. com, p. 74-76, 247, 252, 286.

[25] Ramachandran, Vilayanur S and Hubbard, Edward (2003), "Hearing Colors, Tasting Shapes", *Scientific American*, April 13.

[26] Ramachandran, Vilayanur S., and Hubbard, Edward (2001), "Synesthesia – A Window into Perception, Thought, and Language", *Journal of Consciousness Studies, No. 12*, p. 17.

[27] Pisano, Richard (2012), "Dear Dr. Grinspoon", in Grinspoon, Lester (ed.) *http://marijuana-uses.com/dear-mom-and-dad-by-rob/* Accessed April 20, 2022.

[28] Novak, William (1980), *High Culture. Marijuana in the Lives of Americans*. The Cannabis Institute of America, Massachusetts, p. 97.

[29] Baudelaire, Charles, *Les Paradis Artificiels* (1966), Garnier-Flammarion, p. 47.

[30] See for instance Kowal MA, Hazekamp A, Colzato LS, van Steenbergen H, van der Wee NJ, Durieux J, Manai M, Hommel B. (2015) "Cannabis and creativity: highly potent cannabis impairs divergent thinking in regular cannabis users." *Psychopharmacology* (Berl). 2015 Mar; 232(6):1123-34. doi: 10.1007/s00213-014-3749-1. Epub 2014 Oct 7, PMID: 25288512; PMCID: PMC4336648.

[31] Sternberg, Robert J. (2011), "Creativity". *Cognitive Psychology* (6 ed.). Cengage Learning. p. 479. ISBN 978-1-133-38701-5.

[32] At least I did so, this is actually a situation I experienced with a Chinese Thai Chi grandmaster more than 30 years ago.

[33] Colzato, Lorenza & Hommel, Bernhard & Beste, Christian, (2020). "The Downsides of Cognitive Enhancement." *The Neuroscientist. 27.* *10.1177/1073858420945971.*

[34] Wallas, Graham (1926), *The Art of Thought*, https://archive.org/details/theartofthought. Accessed March 10, 2022.

[35] Wallas, Graham (1926), *The Art of Thought*, Chapter 4, p. 81, https://archive.org/details/theartofthought. Accessed March 10, 2022.

[36] Note that I do not suggest a high has to be bad, generally, for playing basketball. I am sure there are users who feel just fine playing basketball while they are high.

[37] Marincolo, Sebastián (2021), *The Art of the High. Your Guide to Using Cannabis for an Outstanding Life*, Self-Published. *ISBN-13: 978-3981771220.*

[38] See e.g. Sandkühler S, Bhattacharya J. (2008), "Deconstructing insight: EEG correlates of insightful problem solving." *PLoS One. 2008 Jan 23;3(1):e1459. doi: 10.1371/journal.pone.0001459. PMID: 18213368; PMCID: PMC2180197.*

[39] Davis, M. A. (2009), "Understanding the relationship between mood and creativity: A meta-analysis. Organizational Behavior and Human Decision Processes," 108(1), 25e38.https://doi.org/10.1016/j.obhdp.2008.04.0.

[40] See for instance: Schafer G, Feilding A, Morgan CJ, Agathangelou M, Freeman TP, Valerie Curran H. (2012), "Investigating the interaction between schizotypy, divergent thinking and cannabis use." *Conscious Cogn. 2012* Mar; 21(1):292-8. doi: 10.1016/j.concog.2011.11.009. Epub 2012 Jan 9. PMID: 22230356; PMCID: PMC3657189.

[41] Also, see: Kowal MA, Hazekamp A, Colzato LS, van Steenbergen H, van der Wee NJ, Durieux J, Manai M, Hommel B. (2015), "Cannabis and creativity: highly potent cannabis impairs divergent thinking in regular cannabis users." *Psychopharmacology* (Berl). 2015 Mar; 232(6):1123-34. doi: 10.1007/s00213-014-3749-1. Epub 2014 Oct 7. PMID: 25288512; PMCID: PMC4336648.

[42] Luisa Prochazkova, Bernhard Hommel (2020), Chapter 6, p.124 from: *Creativity and the Wandering Mind. https:// doi.org/10.1016/B978-0-12-816400-6.00006-7121©2020 Elsevier Inc.*

[43] Chun, Charlotte, and Hupe, Jean-Michel (2013), "Are synesthetes different beyond their synesthetic associations?" *Journal of Vision*, July 2013, Issue 9.

[44] Sternberg, R. J. (2020), "What's wrong with creativity testing?," *The Journal of Creative Behavior*, 54(1), 20–36. https://doi.org/10.1002/jocb.237

[45] Marincolo, Sebastián (2021), *The Art of the High. Your Guide to Using for an Outstanding Life.* Self-published, ISBN-13: 978-3981771220.

[46] Piomelli, D., & Russo, E. B. (2016), "The Cannabis sativa Versus Cannabis indica Debate: An Interview with Ethan Russo, MD." *Cannabis and cannabinoid research*, 1(1), 44–46. https://doi.org/10.1089/can.2015.29003.ebr.

[47] Geirland, John (1996), "Go With The Flow". *Wired*, September, Issue 4.09.

[48] For an interesting approach for cannabis users which suggests using creative techniques combined with a high, I recommend, "The Creative Thinking Journal," *https://www. pilgrimsoul.com/journals.*

[49] Marincolo, Sebastián (2021), *The Art of the High. Your Guide to Using for an Outstanding Life.* Self-published, ISBN-13: 978-3981771220

[50] For a brilliant collection of pot VIP's and what we know about their cannabis use, see Ellen Komp's *https://www.veryimportantpotheads.com.*

[51] Compare Komp, Ellen, https://www.veryimportantpotheads.com/groucho.html. Accessed March 18, 2022.

[52] I have done this for Walter Benjamin and Carl Sagan in my essay collection: Sebastián Marincolo (2015), *What Hashish Did to Walter Benjamin. Mind-Altering Essays on Marijuana.* Khargala Press, Stuttgart.

[53] Mill, John Stuart (1859/1975) "On Liberty". Edited by David Spitz. Toronto: W. W. Norton. Numbers in parentheses refer to the page number of Mill's essay in this 1975 printing.

[54] Boire, Richard Glen (2002), "John Stuart Mill & the Liberty of Inebriation*", The Independent Review*, Vol. VII, No. 2. Fall 2002.

[55] Mill, John Stuart (1859/1975), *On Liberty.* Edited by David Spitz. Toronto: W. W. Norton.

[56] Sagan, Carl/ "Mr. X", (1971), in: Grinspoon, Lester, *Marihuana Reconsidered*, Harvard University Press, p. 127.

[57] Ibid., p. 128.

[58] Mailer, Norman (1979), Interview with *High Times Magazine*, Issue #49, 1979.

[59] Miller, A.I. (1975), "Albert Einstein and Max Wertheimer: A Gestalt Psychologist's View of the Genesis of Special Relativity Theory", *History of Science*, Vol. 13, p.75-103.

[60] For a good compilation see Sternberg, Robert and Davidson, Janet E. (eds.) (1995), *The Nature of Insight,* Cambridge, Massachusetts: A Bradford Book, The MIT Press.

[61] Yawger, N.S. (1938), "Marijuana: Our New Addiction." *American Med. Sci.*, 195, p. 353.

[62] Novak, William (1980), *High Culture: Marijuana in the Lives of Americans*. Massachusetts: The Cannabis Institute of America, Inc., p. 97.

[63] Marincolo, Sebastián, "Carl Sagan, Cannabis, and the Right Brain Hemisphere", in: Marincolo, Sebastián (2015), *What Hashish Did To Walter Benjamin. Mind-Altering Essays on Marijuana*, Khargala Press, Stuttgart.

[64] Tart, Charles T. loc. cit., p. 133.

[65] Brady, Pete, "Marijuana as an Enhancer of Music Therapy", in: *marijuana-uses.com,* Grinspoon, Lester (ed.). Accessed March 15, 2022.

[66] Twinkly (anonymous), "Twinkly", in: *marijuana-uses. com*, Grinspoon, Lester (ed.). Accessed March 13, 2022.

[67] Tart, Charles T. loc. cit., p. 109-126.

[68] Damasio, Antonio (2006), *Looking for Spinoza. Joy, Sorrow, and the Feeling Brain*. Harcourt Books, p. 106.

[69] Blakeslee, Sandra, and Blakeslee, Matthew (2008), *The Body has a Mind of Its Own*, Random House, p. 5.

[70] Novak, William (1980). *High Culture: Marijuana in the Lives of Americans*. Massachusetts: The Cannabis Institute of America, Inc., p. 138-9.

[71] Byrne, Jon (2012), "Marijuana Stimulates Creativity and Enriches Experience," in: Grinspoon, Lester (ed.) marijuana-uses.com, Grinspoon, Lester (ed.).

[72] Goldfarb, Aaron (2017). "Wine Slang 101: How to Talk Like a Sommelier", *https://firstwefeast.com/drink/ wine-vocabulary-sommelier*, First We Feast, Feb 8, 2017, Accessed March 30, 2022.

[73] For the distinction of the components of human empathic understanding, see Molenbergh, Pascal (2017), "Understanding others' feelings: what is empathy and

why do we need it?", *https://theconversation.com/ understanding-others-feelings-what-is-empathy-and-why-do-we-need-it-68494.*

[74] Truly a sad evolutionary image that reminds me of one of the most important movies coming out of Hollywood, John Huston's *Misfits* (1960), in which Clark Gable and Montgomery Clift are trying to uphold their "free existence" as cowboys by catching wild horses, who are then processed into dog food. Marilyn Monroe, in her last role, as did Clark Gable, protested out of compassion for the horses.

[75] Tart, Charles T., loc. cit., p. 133.

[76] Ibid.

[77] Novak, William (1980), *High Culture: Marijuana in the Lives of Americans.* Massachusetts: The Cannabis Institute of America, Inc., p. 90.

[78] *Ibid.,* p. 87.

[79] "Steven" (anonymous) (2012), "What Marijuana Has Done for Me," in: Grinspoon, Lester (ed.). *marijuana-uses. com.* Accessed March 20, 2022.

[80] See for instance, Goldman, Alvin I. (2006), *Simulating Minds. The Philosophy, Psychology, and Neuroscience of Mindreading.* New York: Oxford University Press.

[81] See for instance Iacoboni, M. (2008), *Mirroring people: The new science of how we connect with others.* Farrar, Straus and Giroux.

[82] Ramachandran, Vilayanur Mirror (2000) "Neurons and imitation learning as the driving force behind 'the great leap forward' in human evolution", *The Edge*, January 6, 2000, https://www.edge.org/3rd_culture/ramachandran/ramachandran_index.html , Accessed April 20, 2022

[83] Ramachandran, Vilayanur S; Oberman, Lindsay M (2006), *"Broken Mirrors: A Theory of Autism".*

Scientific American. 295 (5): 62–9. doi:10.1038/ scientificamerican0607-20sp. PMID 17076085. Accessed April 20, 2022.

[84] Heyes, Cecilia and Catmur, Caroline, "What Happened to Mirror Neurons?", *Perspectives on Psychological Science Volume 17, Issue 1*, January 2022, p.153.

[85] Andreou M, Skrimpa V., "Theory of Mind Deficits and Neurophysiological Operations in Autism Spectrum Disorders: A Review." *Brain Sci. 2020 Jun 20*;10(6):393. doi: 10.3390/brainsci10060393. PMID: 32575672; PMCID: PMC7349236.

[86] Rebecca Brewer, Jennifer Murphy (2016), "People with autism can read emotions, feel empathy," https://www. spectrumnews.org/author/rebeccabrewer.

[87] Anonymous (2012), "Marijuana and its Meaning for Me", in: *marijuana-uses.com*, Grinspoon, Lester (ed.). Accessed March 15, 2022.

[88] Marie Myung-Ok Lee, "Why I give Pot to my 9-year old son", https://web.archive.org/web/20100109103204/http:// www.doublex.com/section/health-science/why-i-give-my-9-year-old-pot. Accessed April 20, 2022.

[89] Ibid.

[90] See for instance: Interview with Goldstein, Bonnie (2022), "Research Breakthrough: Cannabis & Autism". Conducted by Lee, Martin A., Jan 25, 2022, *https://www.projectcbd.org/ medicine/research-breakthrough-cannabis-autism.* Accessed March 29, 2022.

[91] This is an edited and updated version of my essay, "Cannabis and – What a High Can Do For You" (2016), which first appeared for the expert blog of Sensi Seeds Amsterdam, https://sensiseeds.com/en/blog/cannabis-and-sex-what-a-high-can-do-for-your-libido/" (updated blog version from 2020). Accessed April 20, 2022.

[92] Timothy (anonym), "Dear Honey" (2012), in: Grinspoon,

Lester (ed.), *marijuana-uses.com*. Accessed April 15, 2022.

[93] See Marincolo, Sebastián (2015), "The Effects of Marijuana on Body Image Perception," in: Marincolo, Sebastián, *What Hashish Did To Walter Benjamin Mind-Altering Essays on Marijuana*. Khargala Press, Stuttgart.

[94] Carl Sagan, (1971), "Mr. X", in: Grinspoon, Lester, *Marijuana Reconsidered*, Harvard University Press, p.123-130.

[95] Burton, Richard F. (2013), pp. 92-93, *The Book of the Thousand Nights and a Night* (Vol. 3), London: Forgotten Books. (Original work published 1894)

[96] Cleaves, E. (Anonymous), "We're Not Bluffing Anymore", in: Grinspoon, Lester (ed.), http://marijuana-uses.com/were-not-bluffing-anymore-by-e-cleaves/. Accessed April 20, 2022.

[97] This conclusion is at least in parts confirmed by a recent survey, see Wiebe E, Just A. (2019), "How Cannabis Alters Sexual Experience: A Survey of Men and Women." *J Sex Med. 2019 Nov;16(11):1758-1762. doi: 10.1016/j. jsxm.2019.07.023. Epub 2019 Aug 22. PMID: 31447385.*

[98] Compare Androvicova R, Horacek J, Stark T, Drago F, Micale V., "Endocannabinoid system in sexual motivational processes: Is it a novel therapeutic horizon?" *Pharmacol Res. 2017 Jan;115:200-208. doi: 10.1016/j.phrs.2016.11.021. Epub 2016 Nov 21. PMID: 27884725.*

[99] Michael R. Aldrich (1977), "Tantric Cannabis Use in India", *Journal of Psychedelic Drugs*, 9:3, 227-233, doi: 10.1080/02791072.1977.10472053, p. 227.

[100] Shah, N.C. (2014), "The Discovery & Mystery of Soma Plant and its Identification", *The Scitech Journal*, Volume 01, Issue 11.

[101] Keith, A.B. (1925), "Religion and Philosophy of the Veda and Upanishads, Pt 1", Harvard Oriental Series 31: Cambridge, Mass.: Harvard University Press, p.168.

[102] Michael R. Aldrich (1977), "Tantric Cannabis Use in

India", *Journal of Psychedelic Drugs*, 9:3, 227-233, DOI:
10.1080/02791072.1977.10472053, p. 227.

[103] Flood, Gavin (1996), *An Introduction to Hinduism*.
Cambridge: Cambridge University Press.

[104] Doninger O'Flaherty, Wendy (1981), "Śiva: The Erotic
Ascetic," page 84.

[105] Abel, E.L. (1980). *The First Twelve Thousand Years*. New
York: McGraw Hill.

[106] Rätsch, Christian, (2014), *https://www.youtube.com/
watch?v=dYeYo1sFAXY&t=908s* (In German).

[107] Wilson, Anton Robert (1989/2020), *Ishtar Rising: Why
the Goddess went to Hell & What to Expect now that She's
Returning*, Second Edition, Hilaritas Press, Grand Junction,
p.12.

[108] Michael R. Aldrich (1977), "Tantric Cannabis Use in
India", *Journal of Psychedelic Drugs, 9:3, 227-233, DOI:
10.1080/02791072.1977.10472053*, p. 227.

[109] Ibid.

[110] Ibid., p. 231.

[111] Ibid., p. 231.

[112] See for instance Marincolo, Sebastián (2010*), High:
Insights on Marijuana*, Dogear Publishing.

[113] Aldrich, Michael R. (1977), "Tantric Cannabis Use in
India", *Journal of Psychedelic Drugs, 9:3, 227-233, DOI:
10.1080/02791072.1977.10472053, p. 231.*

[114] Ibid, p. 231.

[115] Alger, Bradley E. (2013), "Getting High on the
Endocannabinoid System", Cerebrum. 2013 Nov-Dec; 2013:
14. *Published online 2013 Nov 1. PMCID:* PMC3997295
PMID: *24765232.*

[116] See V. F. Kitchigina (2021), "Cannabinoids, the Endocannabinoid System, and Cognitive Functions: Enemies or Friends?", *Neuroscience and Behavioral Physiology 2021, 51 (7) , 893-914. https://doi.org/10.1007/s11055-021-01148-5.*

[117] McPartland, JM (2018), "Cannabis systematics at the levels of family, genus, and species", *Cannabis and Cannabinoid Research 3:1, 203–212, doi: 10.1089/ can.2018.0039.*

[118] Kilaru, Aruna, Chapman, Kent. D. (2020), "The endocannabinoid system", Essays in Biochemistry EBC20190086 *https://doi.org/10.1042/EBC20190086,* p. 9.

[119] Dudok, B, and Soltesz, I. (2022), "Imaging the endocannabinoid signaling system," *J Neurosci Methods.* 2022 Feb 1;367:109451. *https://doi: 10.1016/j. jneumeth.2021.109451* Epub 2021 Dec 15. PMID: 34921843; PMCID: PMC8734437.

[120] Marsicano G, Wotjak CT, Azad SC, Bisogno T, Rammes G, Cascio MG, Hermann H, Tang J, Hofmann C, Zieglgänsberger W, Di Marzo V, Lutz B. "The endogenous cannabinoid system controls extinction of aversive memories," *Nature. 2002 Aug 1;418(6897):530-4. doi: 10.1038/nature00839. PMID: 12152079.*

[121] For a collection of some scientific studies see Marijuana Policy Project, "PTSD and Medical Cannabis Programs', https://www.mpp.org/issues/medical-marijuana/ptsd-medical-cannabis-programs/ Accessed March 16, 2022.

[122] Oreskes, Naomi, and. Conway, Erik M. (2012) Merchants of Doubt: How a Handful of Scientists Obscured the Truth on Issues from Tobacco Smoke to Global Warming. Bloomsbury Paperbacks; UK ed. edition

[123] Sun, Lilias, (March 7, 2014), "HMS's Grinspoon Calls for NFL Funding for Cannabis Research," *The Harvard Crimson,* https://www.thecrimson.com/article/2014/3/7/HMS-professor-NFL-cannabis-research/. Accessed March 15th, 2022.

[124] Kitchigina, Valentina (2021), "Cannabinoids, the Endocannabinoid System, and Cognitive Functions: Enemies or Friends?" *Neuroscience and Behavioral Physiology. 51. 10.1007/s11055-021-01148-5, p. 905-906.*

[125] Russo, E. B. (2016), "Clinical Endocannabinoid Deficiency Reconsidered: Current Research Supports the Theory in Migraine, Fibromyalgia, Irritable Bowel, and Other Treatment-Resistant Syndromes." *Cannabis and cannabinoid research, 1(1), 154–165. https://doi.org/10.1089/can.2016.0009.*

[126] Cogan, Peter S. (2020), "Practical Considerations of Hypotheses and Evidence in Cannabis Pharmacotherapy: Refining Expectations of Clinical Endocannabinoid Deficiency", *Journal of Dietary Supplements, 17:5, 608-624, doi: 10.1080/19390211.2020.1769246.*

[127] Ibid, p. 895-896.

[128] For an interesting perspective on how the brain achieves this balance with a labor division between two brain hemispheres which process information fundamentally differently, see *McGilchrist, Iain (2009), The Master and His Emissary: The Divided Brain and the Making of the Western World.* USA: Yale University Press. ISBN 978-0-300-14878-7.

[129] Compare Marincolo, Sebastián (2021), *The Art of the High: Your Guide to Using Cannabis for an Outstanding Life*, Self-Published. *ISBN-13: 978-3981771220.*

[130] Sontag, Susan (1977), *On Photography*, Penguin Books, London, p. 24.

[131] Felman, Adam (2021), "What is Addiction", *https://www.medicalnewstoday.com/articles/323465,* Accessed March 15, 2022.

[132] Maté, Gabor (2009), *In the Realm of Hungry Ghosts: Close Encounters with Addiction*, "Chapter 2, The Lethal Hold of Drugs." Knopf Canada

[133] Ibid, p.33.

[134] Sadly, I do not recall who gave the talk. I was there only in my function as a documentary photographer.

[135] *https://www.mentalhealth.org.uk/sites/default/files/truth_ about_self-harm_NEW_BRAND_0.pdf*

[136] See for instance "Medical Pot May Help You Avoid Opioids for Back Pain, Arthritis" *https://www.webmd.com/back-pain/news/20220322/ arthritis-back-pain-medical-pot-may-help-you-avoid-opioid- painkillers?src=RSS_PUBLIC*. Accessed March 28, 2022.

[137] Hill, M., Campolongo, P., Yehuda, R. et al. (2018), "Integrating Endocannabinoid Signaling and Cannabinoids into the Biology and Treatment of Posttraumatic Stress Disorder.", *Neuropsychopharmacol. 43, 80–102. https://doi. org/10.1038/npp.2017.162.*

[138] See for instance: Vyas MB, LeBaron VT, Gilson AM. "The use of cannabis in response to the opioid crisis: A review of the literature." *Nurs Outlook. 2018 Jan-Feb;66(1):56-65. doi: 10.1016/j.outlook.2017.08.012. Epub 2017 Sep 21. PMID: 28993073.* For an overview of studies on the cannabinoid CBD for Alcohol Use Disorder, see Coelho, Steph (2021) "CBD for Alcohol Use Disorder: Can Taking CBD Help Decrease Alcohol Consumption?", *https://www.healthline. com/health/cbd-for-alcoholism.* For a possible treatment of methamphetamine abuse, see Razavi, Y., Keyhanfar, F., Shabani, R., Haghparast, A., Mehdizadeh, M. (2021). "Therapeutic Effects of Cannabidiol on Methamphetamine Abuse: A Review of Preclinical Study." *Iranian Journal of Pharmaceutical Research*, 20(4), 152-164. *https://doi: 10.22037/ ijpr.2021.114918.15106.*

[139] Compare Nutt, David J., King, Leslie A., Phillips, Lawrence D. (2010), "Drug harms in the UK: a multicriteria decision analysis." In: *The Lancet; DOI: 10.1016/ S0140-6736(10) 61462-6.* Also, compare Deutsche Hauptstelle für Suchtfragen, *Jahrbuch Sucht 2019.*

[140] National Academies of Sciences, Engineering, and Medicine. 2017. *"The Health Effects of Cannabis and Cannabinoids: The Current State of Evidence and Recommendations for Research."* Washington, DC: The National Academies Press. *https://doi.org/10.17226/24625.*

[141] Compare Grinspoon, Lester, (1971*), Marijuana Reconsidered.* Bantam Books, p.16.

[142] Fossier, A.E., (1931), "The Marihuana Menace*", New Orleans Med. Surg. J., 84*, p.249.

[143] Bonnie, Richard J., and Whitebread II, Charles H., *The Marihuana Conviction: A History of Marihuana Prohibition in the United States* (Charlottesville, VA: University Press of Virginia, 1974), 35, 52.

[144] See Campos, Issac (2018), "Mexicans and the Origins of Marijuana Prohibition in the United States: A Reassessment." *Social History of Alcohol and Drugs, Volume 32.*

[145] Lee, Martin A. (2012), Smoke Signals. A Social History of Marijuana – Medical, Recreational, and Scientific. Scribner, New York.

[146] Woodward, C. Vann and McFeely, William S. (2001), *The Strange Career of Jim Crow*, Oxford University Press, p. 7.

[147] Seguin, Charles; Rigby, David (2019), "National Crimes: A New National Data Set of Lynchings in the United States, 1883 to 1941". *Socius: Sociological Research for a Dynamic World. 5. https://doi:10.1177/2378023119841780.* ISSN 2378-0231. S2CID 164388036.

[148] Grinspoon, Lester, (1971*), Marijuana Reconsidered.* Bantam Books, p. 27f.

[149] U.S. Congress, house of Representatives, Congressional Record, 75th Congress, 1st session, June 14, 1937, p. 5575.

[150] Mayor LaGuardia's Committee on *Marijuana* (1944), "The Marijuana Problem in the City of New York," (Lancaster , Pa . : Jacque Cattell Press; Metuchen , N . J .: Scarecrow Reprint Corporation, 1973.

[151] Lee, Martin A. (2012), *Smoke Signals: A Social History of Marijuana – Medical, Recreational, and Scientific*. Scribner, New York, p. 62.

[152] Hay, Mark (2018), "In Istanbul, Drinking Coffee in Public Was Once Punishable by Death", *https://www.atlasobscura. com/articles/was-coffee-ever-illegal Accessed April 20, 2022.*

[153] Ibid.

[154] Beiglböck, W. (2016), *Koffein: Genussmittel oder Suchtmittel?* Springer Berlin Heidelberg, p.28.

[155] Kohn, Brigitte (01/21/2019), "Friedrich der Große verbietet Kaffee-Rösten", https://www.br.de/radio/bayern2/ sendungen/kalenderblatt/Friedrich-der-grosse-erlaesst-kaffeeroestverbot-kaffee-100.html

[156] Köpke, Monika (1/21/2006), *Bier statt Kaffee, https:// www.deutschlandfunk.de/bier-statt-kaffee-100.html*

[157] Becker, Howard S. (1963), *Outsiders: Studies in the Sociology of Deviance*. (New York: The Free Press).

[158] Carole Cadwalladr and Emma Graham-Harrison (2018), "Revealed: 50 million Facebook profiles harvested for Cambridge Analytica in major data breach." *https://www.theguardian.com/news/2018/mar/17/cambridge-analytica-facebook-influence-us-election*. Accessed March 14, 2022.

[159] Grinspoon, Lester, (1971) *Marijuana Reconsidered*. A Bantam Book, p.15.

[160] For a great analysis of how Republican think tanks managed to shape debates by creating slogans and metaphors, see Lakoff, George (2004), *Don't Think of an Elephant: Know Your Values and Frame the Debate*. Chelsea Green Publishing.

[161] Compare Marincolo, Sebastián "The Most Powerful Drug Used by Mankind", in: *What Hashish Did To Walter Benjamin. Mind-Altering Essays on Marijuana*. (2015), Khargala Press, Stuttgart, p.146.

[162] Baum, Dan (2016), "Legalize it all. How to win the war on drugs.", *Harper's Magazine https://harpers.org/archive/2016/04/legalize-it-all/* Accessed March 20, 2022.

[163] Zimmer, Lynn, and Morgan, John P. (1997), *Marihuana Myths, Marihuana Facts. A Review of the Scientific Facts.* The Lindesmiths Center: New York.

[164] Nutt, David, and Moss, Brigid (2021), *Cannabis (seeing through the smoke): The New Science of Cannabis and Your Health.* eBook ISBN 978 1 529 36050 9, Yellow Kite, London.

[165] See Marincolo, Sebastián (2017), Marijuana, Escapism, and Mind Travelling. *https://sensiseeds.com/en/blog/marijuana-escapism-and-mind-traveling/* Accessed March 14th, 2022.

[166] See "Marihuana Prohibition Has Not Curtailed Marihuana Use by Adolescents." (2014), from the Marijuana Policy Project, *http://www.ukcia.org/research/young/adolescent.php* Accessed April 20, 2022.

[167] See Marijuana Policy Project (MPP), "Marijuana Prohibition Facts", *https://www.mpp.org/issues/legalization/marijuana-prohibition-facts/* Accessed March 15th, 2022.

[168] See EMCDDA, "Cannabis use in the last year in Europe – young adults (15-34)", *https://www.emcdda.europa.eu/media-library/cannabis-use-last-year-europe-%E2%80%93-young-adults-15-34_en*

[169] Gabri AC, Galanti MR, Orsini N, Magnusson C (2022), "Changes in cannabis policy and prevalence of recreational cannabis use among adolescents and young adults in Europe—An interrupted time-series analysis." *PLOS ONE 17(1): e0261885. https://doi.org/10.1371/journal.pone.0261885*

[170] Compare: Anderson DM, Sabia JJ, "Notice of Retraction and Replacement. Anderson et al. Association of Marijuana Legalization With Marijuana Use Among US High School Students, 1993-2019." *JAMA Netw Open. 2021;4(9):e2124638. JAMA Netw Open. 2022;5(3):e221473.* doi:10.1001/jamanetworkopen.2022.1473. See, also: Marijuana Policy Project, (3/10, 2022) "Teen Marijuana Use Does Not Increase Following Marijuana Policy Reforms", *https://www.mpp.org/issues/legalization/teen-marijuana-use-does-not-increase/.*

[171] Marincolo, Sebastián (2016/updated 2020), "Marijuana, Escapism, and Mind Traveling", blog post, *https://sensiseeds.com/en/blog/marijuana-escapism-and-mind-traveling/* Accessed April 20, 2022.

[172] Siegel Ronald K. (2005), *Intoxication: The Universal Drive for Mind-Altering Substances.* Park Street Press, p. vi.

Opening Photograph

Macro shot of trichomes on a cannabis leaf.

Many of the photos in this book belong to
Sebastián Marincolo's large framed limited
photo art edition *The Art of Cannabis* and can
be obtained through his website:
www.sebastianmarincolo.de.

Closing Photograph

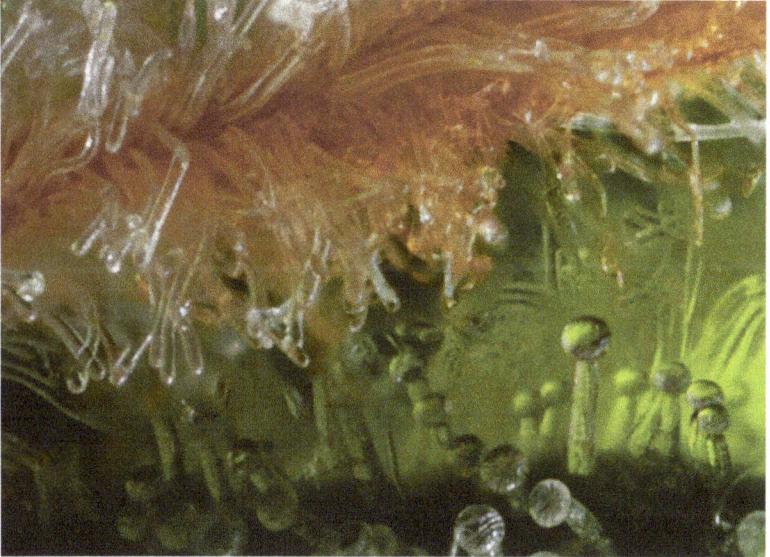

Trichomes and a ripe stigma on a cannabis flower. When the stigmas ripen, they turn from a greenish crystal white to amber, red, pink, or other colors, depending on the variety of the cannabis plant.

HILARITAS
PRESS

Publishing the Books of Robert Anton Wilson
and Other Adventurous Thinkers

www.hilaritaspress.com

www.ingramcontent.com/pod-product-compliance
Lightning Source LLC
Chambersburg PA
CBHW052110030426
42335CB00025B/2912